FIREFLY

GUIDE TO
WETLANDS

DR. PATRICK DUGAN
GENERAL EDITOR

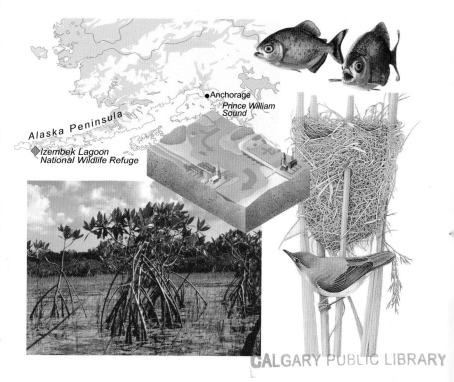

Anchorage
Prince William Sound

Alaska Peninsula

Izembek Lagoon
National Wildlife Refuge

FIREFLY BOOKS

A FIREFLY BOOK

Published by Firefly Books Ltd. 2005

Copyright © 2005 Philip's

First printing

Publisher Cataloging-in-Publication Data (U.S.)

Guide to wetlands / Patrick Dugan, general editor.
[304] p. : col. photos.; cm.
Includes index.
Summary: A Guide to the wetlands of the world and how they function as environments and habitats for wildlife and humans.
ISBN 1-55407-111-9 (pbk.)
1. Wetlands. 2. Wetland ecology. I. Dugan, Patrick, 1955- II. Title.
333.91/8 dc22   QH87.3D84 2005

Library and Archives Canada Cataloguing in Publication

Guide to wetlands / Patrick Dugan, general editor.
Includes index. ISBN 1-55407-111-9
1. Wetlands. 2. Wetland ecology. I. Dugan, Patrick, 1955-
QH541.5.M3G84 2005   578.768   C2005-902452-6

Published in the United States by
Firefly Books (U.S.) Inc.
P.O. Box 1338, Ellicott Station
Buffalo, New York 14205

Published in Canada by
Firefly Books Ltd.
66 Leek Crescent
Richmond Hill, Ontario L4B 1H1

COMMISSIONING EDITOR *Frances Button*
EDITOR *Joanna Potts*
EXECUTIVE ART EDITOR *Mike Brown*
DESIGNER *Caroline Ohara*
PRODUCTION *Åsa Sonden*

FRONT COVER : *tl* Western Canada and Alaska map, Andrew Thompson, Philip's; *tr* Serrasalmus nattereri, Philip's; *c* floodplains diagram, Colin Rose; *bl* red mangroves, Gary M. Stolz/USFWS; *br* reed warbler, Philip's
BACK COVER : *bl* cockle, Philip's; *br* green tree frog, Jane M. Rohling/USFWS
THIS PAGE : Great blue heron, Philip's
CONTENTS PAGE (from top to bottom) : arctic fox, Keith Morehouse/USFWS; Gurra Gurra Wetlands, the Australian Government, Department of Environment and Heritage Collection, reproduced by permission. Taken by Merran Williams; the Lague de Porto-Novo, Nassima Aghanim, *Ramsar;* Lacassine National Wildlife Refuge, John and Karen Hollingsworth/USFWS

Printed in China

CONTRIBUTORS
Eastern Canada and Greenland David Boertmann, Steffen Brogger Jensen, Clayton Rubec
Western Canada and Alaska Tom Dahl, Clayton Rubec, Jim Thorsell
The United States - The Lower "Forty-Eight" Tom Dahl, Joe Larson, Dan Scheidt
Mexico, Central America and the Caribbean Alejandro Yanez Arancibia, Peter Bacon, Monica Herzig, Enrique Lahmann
Northern South America and the Amazon Basin Antonio Diegues, Stefan Gorzula, Francisco Rilla
Southern South America Argentino Boneto, Pablo Canevari, Maria Marconi, Victor Pullido, Francisco Rilla, Hernan Verascheure, Yerko Vilina
Northern Europe David Boertmann, Steinar Eldoy, Jens Enemark, Palle Uhd Jepsen, Esko Joutsamo, Torsten Larsson, Karsten Laussen, Hans Meltofte, Ole Thorup
West and Central Europe Andrew Craven, Nic Davidson, P Gatescu, Liz Hopkins, Zbig Karpowicz, Edward Maltby, Francois Sarano, A Vadineanu, Jurgen Voltz, Edith Wenger
The Mediterranean Basin George Catsadorakis, Alain Crivelli, A Gerakis, Alain Johnson, Thymio Papayannis, Jamie Skinner, Nergis Yazgan
The Middle East Andrew Price, Derek Scott
East Africa and the Nile Basin Geoff Howard, Paul Mafabi, Steven Njuguna, MA Zahran
West and Central Africa Pierre Campredon, Jean-Yves Pirot, Ibrahima Thiaw
Southern Africa Geoff Howard, Tabeth Matiza, Rob Simmons
Northern Asia Genady Golubev, Vitaly G Krivenko, Mike Smart, Vadim G Vinogradov
Central and South Asia Zakir Hussain, Peter-John Meynell, Sam Samarakoon
East Asia Derek Scott, Satoshi Kobayashi
Southeast Asia Zakir Hussain, Duncan Parish, Marcel Silvius
Australia Jim Davie, Max Finlayson, Jim Puckeridge
New Zealand and the Pacific Department of Conservation (New Zealand), Derek Scott

Wetlands maps produced by Andrew Thompson using data from UNEP-World Conservation Monitoring Centre

Contents

What are wetlands?

Most of us are familiar with wetlands in some shape or form. A nearby pond, the trout stream or the local estuary, for example, are just three of the many types of wetland that are widespread throughout temperate regions. But further south, in tropical and subtropical regions, there are muddy tidal flats, expansive flood-plains and misty swamplands: three very different environments with very different plants and animals, but these are wetlands too.

Wetland diversity

There are more than 50 definitions of wetlands in use throughout the world. Among these the broadest, and therefore that which is used most widely on an international scale, is provided by the *Ramsar Convention on Wetlands of International Importance, Especially as Waterfowl Habitat*. Ramsar is an Iranian city lying on the shores of the Caspian Sea, and it was here that the Wetland Convention was adopted in 1971. Designed to

▼ *Reed boats on Lake Titicaca, the world's highest navigable lake. Situated on the border between Bolivia and Peru, Titicaca's surface is some 12,500 ft (3,810 m) above sea level. Large stretches of the lake's shoreline consist of marsh and reedbeds. Surrounded by high-altitude desert, the lake provides an attractive environment for many species of fish and birds, as well as a population of tens of thousands of people.*

provide international protection to the widest possible group of wetland ecosystems, the Ramsar Convention uses a broad definition of wetlands including swamps and marshes, lakes and rivers, wet grasslands and peatlands, oases, estuaries, deltas and tidal flats, near-shore marine areas, mangroves and coral reefs, and man-made sites such as fish ponds, rice paddies, reservoirs and salt pans.

Estuaries, mangroves and tidal flats

Estuaries form where rivers enter the sea. The daily tidal cycle and the intermediate salinity between salt and freshwater, which are characteristic of these ecosystems, make them difficult places in which to live. However, those species that have adapted successfully thrive in these conditions. Indeed, estuaries and inshore marine waters are among the most naturally fertile habitats in the world.

Estuaries are found in all regions of the world, but their productivity varies with climate, hydrology (the water cycle) and coastal land forms. Many estuaries are associated with important lagoon systems, some of which have been created by the closure of one of the estuaries' outlets to the sea. In temperate regions, intertidal mud and sand flats, salt marshes and scattered, rocky outcrops are common features of estuaries. In the tropics and sub-tropics, however, mangroves dominate many coastal habitats and are characteristic of most estuaries.

Variously referred to as "coastal woodland", "tidal forest" and "mangrove forest", mangroves comprise very diverse plant communities, whose composition varies greatly from region to region. Even within the same delta, the composition of the mangrove community can vary substantially according to the conditions of salinity, tidal system and substrate (the soil foundation). Approximately 80 species of plant are recognized as being mangroves. They share a variety of adaptations that enable them to grow in the unstable conditions of estuarine habitats in the tropics and subtropics.

Although mangroves, mud flats and other coastal wetland habitats are normally most extensive around estuaries, they are also found along areas of open coast. For example, in Mauritania, the Banc d'Arguin, Africa's largest system of tidal flats, receives no significant surface inflow of freshwater. And sandy beaches, which are characteristic of almost every coastal country, support important populations of wildlife, including migratory shorebirds and nesting marine turtles.

Floodplains and deltas

As rivers swell with seasonal rainfall they slowly rise above the river channel and under natural conditions flow out over the neighboring plain. This pattern of seasonal flooding was once a common feature of most of the world's rivers. Today, however, with the increasingly widespread construction of dams and embankments, the natural patterns of flooding have been severely disrupted in many regions. Nevertheless, the annual cycle of inundation and drying of the world's floodplains remains one of the most important forces governing wetland productivity.

Evaporation and transpiration

▶ The various types of wetland shown in the diagram have different hydrological signatures. Estuaries [A] and mangroves [B], for example, depend upon the tides. As the tide changes, the salinity of the water can vary between almost totally saline to freshwater. Animals and plants that inhabit these regions have adapted to survive both the daily flooding and drying out as well as the salinity variation. Floodplains [C] and flooded forests [D] tend to undergo seasonal flooding. Their capacity to store water can be beneficial as they can retain floodwater, preventing flooding downstream, and recharge groundwater. This process of recharging also purifies the water as it slowly filters

through sediment and rock strata. Lakes experience an aging process known as eutrophication. "Old" lakes [E] are characterized by algal growth, a sign of nutrient rich/oxygen poor water. Fertilizer runoff from fields can speed the process. "Young" lakes [F] tend to have clear, oxygen rich/nutrient poor water. Fens [G] and bogs [H] differ

in that fens receive nutrients from groundwater flow and can sustain a wide diversity of plants and animal life. Bogs, on the other hand, receive no groundwater and are therefore very acidic and lacking in nutrients. These harsh conditions are suitable only for plants such as Sphagnum moss, which can tolerate acidic soils.

In many areas of the world, floodplains are found in coastal lowlands and end in estuarine deltas where they become complex mosaics of marine, brackish and freshwater habitats. Alternatively, some of the world's larger rivers spread out over floodplains far inland, many of them covering vast areas that include grassy marshes, flooded forest, oxbow lakes and other depressions. These floodplains are often referred to as inland deltas. Some of the most important floodplains, such as the Inner Niger Delta in Mali, are in arid areas where their exceptional productivity is not only vital to the local economy of the region, but also supports some rare and spectacular concentrations of waterbirds as well as other wildlife.

Freshwater marshes
Stretching from Lake Okeechobee to the southwest tip of Florida, USA, the Everglades covers over 2,700 square miles (7,000 square kilometers). It is one of the world's largest freshwater marshes. At the other extreme of the scale are the marshes that form in small, wet areas wherever groundwater, surface springs, streams or excess runoff cause frequent flooding, or create permanent areas of shallow water.

Although few freshwater marshes can compare in size to the Everglades, the vast number of small marshes makes this type of wetland among the most widespread and important worldwide. Large areas of southern Africa, for example, are dotted with "dambos", small freshwater marshes, which provide essential grazing and agricultural land for many rural communities. In North America, the Prairie Pothole region includes several million freshwater marshes at densities as high as 150 per square mile (60 per square kilometer) in some areas. Some of the larger marshes dominated by papyrus (*Cyperus papyrus*), cattail (*Typha* sp.) and reed (*Phragmites* sp.), and which have standing water throughout most of the year, are normally referred to as swamps rather than wetlands.

G

H

Permeable rock

Groundwater

Impermeable rock

Lakes

The diversity of lakes and ponds is the result of a host of different processes. Some lakes, such as Reelfoot Lake, Tennessee, in the United States, Lake Baikal, in Siberia and Lake Tanganyika, on the borders of the Democratic Republic of Congo, Tanzania, Burundi and Zambia, are formed by folding or faulting of the Earth's crust. Similarly, many crater lakes, including many in the Pacific islands, have been formed through volcanic disturbances. In the Northern Hemisphere, glacial action has been an especially important force. Cirque lakes, thaw lakes, and pothole or kettle lakes all owe their origin to the processes of glacial ice. The action and flow of rivers can also create a variety of different lake types, such as oxbow and alluvial fan lakes, plunge pools and basins. Alpine lakes are formed by landslides and mudflows, while some lakes are remnants of larger ones, formed under more moist prehistoric environments. Shifting sediments by nearshore currents can create shoreline lakes cut off from larger seas of freshwater bodies.

Peatlands

Once thought to be restricted to the high latitudes of the Northern Hemisphere, peatlands are now known to exist on all continents and at all latitudes. They are even found in the tropics, where thick deposits form in association with marsh and swamp, particularly around lake margins and coastal regions. In total, peatlands are estimated to cover some 1.5 million square miles (4 million square kilometers). There is a great diversity of peatland worldwide, the pattern being governed by acidity, climate and hydrology (especially whether the peat is kept wet by direct rainfall or lateral groundwater flow). The highly distinctive, northern wetland landscapes of bog, moor, muskeg and fen are all examples of peatland.

In general terms, peat forms under conditions of low temperature, high acidity, low nutrient supply, waterlogging and oxygen deficiency. These specific circumstances slow the decomposition of dead plant matter. The characteristics of peatland ecosystems, however, are so varied that it is difficult to generalize on their functions and values. For example, some peatlands, namely bogs, are highly acidic and nutrient deficient, others, such as fens, are more or less neutral and rich in nutrients. Peatlands, therefore, include some of the least, as well as some of the most, productive wetlands.

▶ *Sweetwater Marsh National Wildlife Refuge, California, USA. This salt marsh habitat, just 315 acres, is all that's left of the huge salt marshes that once surrounded San Diego Bay. Although not as productive as mangroves, this salt-resistant vegetation forms a vital part of estuaries in northern latitudes. This refuge is home to the endangered least tern (Sterna antillarum), Belding's savannah sparrow (Passerculus sandwichensis belgingi), and light-footed clapper rail (Rallus longirostris levipes).*

Forested wetlands

Swamp forests develop in areas of still water around lake margins and in parts of floodplains, such as oxbow lakes, where water rests for long periods. Their character varies according to geographical location and environment. In the northern United States, for example, red maple (*Acer rubrum*), northern white cedar (*Fraxinus* sp.) and black spruce (*Picea mariana*) are prominent, while in the south, bald cypress (*Taxodium distichum*), black gum (*Nyssa sylvatica*), Atlantic white cedar (*Chamaecyparis thyoides*) and willows (*Salix* sp.) dominate.

In much of Southeast Asia the swamp forests are dominated by paper-bark trees (*Melaleuca* sp.) and other commercially valuable species. In Indonesia, swamp forests of this type cover over 65,000 square miles (170,000 square kilometers). Similar resources are found in the Amazon Basin. Here, the floodplains of the River Amazon and its tributaries support some of the most extensive flooded forests in the world. In Africa, the most extensive areas are found in the Congo Basin, where hundreds of thousands of square kilometers of the flood-plain are densely forested.

How wetlands work

The hydrology of wetlands creates the unique conditions that distinguishes these environments from either terrestrial or deepwater habitats. Hydrological systems of different wetlands vary greatly in terms of frequency of flooding, and duration and depth of waterlogging. A coastal salt marsh typically floods twice a day, while also revolving around a monthly pattern of spring and neap tides. The water level between high and low tides also varies enormously, from being negligible in the case of the Gulf of Mexico or the Mediterranean, to more than 20 feet (6 meters) in many exposed estuaries, and more than 40 feet (12 meters) in special coastal configurations such as those of the Bay of Fundy in Canada. In strong contrast, the water level of many bogs and fens may remain consistently just below the peat surface throughout the year, dropping by a few centimeters or inches only during the summer or a drought period.

A very different pattern occurs on the world's major floodplains. Here, highly seasonal flood peaks give rise to floodwaters which commonly exceed 26 feet (8 meters). The precise pattern and timing of the flood depends on the season and location of rains and the shape of the drainage basin and floodplain.

It is the regularity of the flood pattern, however, which is so important in maintaining the structure and function of wetlands. Without a regular cycle, the productivity of fisheries, vegetation growth cycles and the success of wildlife migrations are seriously affected. In turn, the well-being of many human communities in Africa, Southeast Asia and South and Central America are likewise dependent on the flood pattern for the essential goods and services which the wetlands provide.

Life-bearing water

Water movement is important in the transportation of materials such as sediment, organic matter and nutrients to, from and within wetlands. Water, for example, provides a way for aquatic organisms to travel over what may well be dry land for significant periods of the year. This is vital for migratory fish, such as carp and bass, which have to cross floodplains to spawn in vegetation or swamp forest. And of course rivers themselves provide certain species of fish with the migration pathways they need.

The characteristics of wetlands are influenced strongly by water quality. Waters are classified according to their fertility, which in turn reflects nutrient content – from the

▲ *Algae accumulation, Kesterson National Wildlife Refuge, California, USA. Under these near-stagnant conditions, inflow of nutrients stimulates algal growth, which in turn deoxygenates the water. Although this situation was brought about by a different set of conditions to those in acidic bogs, the result is the same – the water supports few animals or higher plants.*

lowest to the highest: ultraoligotrophic, oligotrophic, mesotrophic, eutrophic and hypereutrophic. These terms apply primarily to lakes, where the extremes are easy to identify. At the oligotrophic end of the scale, waters are very clear, saturated with oxygen, but contain few nutrients essential to plant life, and therefore little animal life. Fast-running mountain streams are good examples of this condition. At the eutrophic extreme of the scale, waters are turbid, support dense algae growths, but have a very low oxygen content and therefore support few animals. Lowland enclosed lakes in agricultural catchments with low turnover of water and high nutrient inputs are characteristic of this condition.

The terminology can be extended generally to the wide range of wetland types. In the case of peatlands, for example, certain types of bogs are oligotrophic. The few nutrients they receive come solely through rainfall. Most fens, however, may be mesotrophic or even eutrophic. Fens usually obtain their nutrients from the lateral flow of groundwater, which carries the dissolved nutrients from soil or rock strata. Thus, while bogs are dependent on a certain amount of rainfall, there is no climatic restriction on the development of fens. The occurrence of fens is based on rock formation and the flow of groundwater.

For this reason, fens exist in regions all over the world where it is too dry for the development of bogs.

Although bogs and fens both comprise accumulations of peat and may have similar water tables, the difference in water chemistry results in different vegetation and habitat for wildlife. Bogs are dominated by plants capable of withstanding acidic conditions, such as *Sphagnum* mosses, cotton grass (*Eriophorum* spp.), sedges like *Carex rostrata* and rushes like *Juncus squarrosus*. Insectivorous plants such as sundew (*Drosera* spp.) are also often present, indicating the need to supplement certain nutrients, particularly nitrogen, in such environments. The nutrient-enriched fens on the other hand support a rich variety of plants, including bulrush, common spike rush, bulbous rush, marsh pennywort and a wide variety of brown mosses.

Succession

Wetlands evolve as a result of natural processes of development. Such change is led by successive waves of varying plant communities, which, as they develop, alter the environment. They produce habitat conditions increasingly less suitable for their own survival, but more

▼ *Tufted sedge (Carex sp.) in Waquoit Bay National Estuarine Research Reserve, Massachusetts, USA. The tussock form of the plant allows it to survive spring flooding. By raising the green parts up on the tussock, the plant can keep growing even at the height of the floods (marked by the transition from dead to live leaves).*

advantageous for other species. This type of succession is known as *autogenic* succession. Another type of succession, in which change is caused by some external factor, such as a warming of the climate, is called *allogenic* succession.

The classic autogenic succession is illustrated in the development of a small freshwater lake or pond. The first flowering plants to colonize the wetland are floating species such as pondweeds (*Potamogeton* sp.) and duckweeds (*Lemna* sp.). Over time, these encourage detritus and sediment to build up around the wetland edges. The shallower water, in turn, allows the emergent

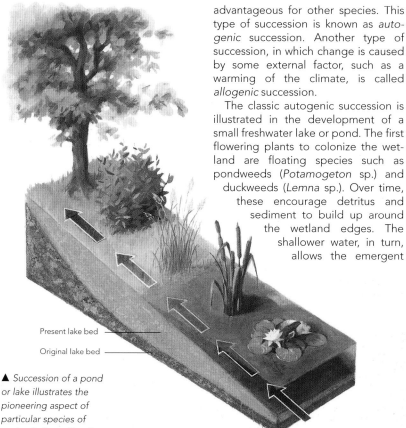

Present lake bed

Original lake bed

▲ *Succession of a pond or lake illustrates the pioneering aspect of particular species of wetland plants. Floating species, such as water lilies, encourage detritus and sediment buildup, which over time (represented by the arrows) enables emergents, such as reeds and rushes, to take root. Further sediment buildup occurs, creating a sufficiently firm substrate on which terrestrial plants can eventually grow.*

species such as reeds and rushes to take root. These trap even more sediment and give way to terrestrial, shrublike plants. In this way, the wetland is transformed to a dry-ground ecosystem by natural development. Succession can be accelerated and extended by human activities such as drainage, river diversion and groundwater abstraction.

Coastal succession

Considerable debate surrounds the existence of succession in coastal wetland systems such as mangroves. Early views were that species such as *Rhizophora* acted as pioneers, trapping sediment and acting as a land builder. This then allowed other mangroves such as *Avicennia* to establish themselves and eventually dominate the space previously occupied by the *Rhizophora*. In time, the pioneer species is forced seaward to survive, and in so doing progressively extends the landward margin. Such theories led to the term "land-builder" to describe the process of accelerated sedimentation and mangrove expansion. Rates of advance of more than 330 feet (100 meters) have been recorded and may lead to spectacular coastline changes. Palenberg, Indonesia, was once a thriving port on the north coast of Sumatra, visited by Marco Polo in 1292. Today, it is 30 miles (50 kilometers) inland.

More recent studies, however, indicate that while mangroves commonly exhibit zonation, there is by no means a uniform pattern and species distribution is determined by much more complex patterns of water level, salinity, sedimentation and flooding regime than previously recognized. So-called pioneers may simply respond to patterns and rates of sedimentation in the tropical coastal zone rather than control it. It may be much more accurate to consider mangroves as land consolidators or protectors rather than creators.

▼ *Yellow water lilies (Nymphaea mexicana) growing on the Mississippi River, USA. Although it is classed as a floating plant, the water lily is in fact firmly rooted. Encased in the bottom mud, a thick rootstock sends stalks up to the surface which terminate either in circular leaves or a single flower. The family Nymphaeaceae have a worldwide distribution, and include the giant water lily (Victoria regia) with leaves that measure up to 6 ft) in diameter.*

▼ *Reed warbler*
(Acrocephalus
scirpaceus) *and nest.*
The reed warbler is a
common bird of reed
beds throughout
Europe, with particularly
high concentrations in
Romania. The reed
warbler's nest is built
round dry reed stems
that safely secure it
above the wet reed
bed. Built in this way,
the reed warbler's nest
remains safe from any
sudden water level
fluctuations.

Wetlands and time

In terms of geological timescales, wetlands retain specific physical features for very short periods. This may be because they are associated with river valleys, flood-plains or deltas, systems which are forever evolving new shapes or structures. Natural development or changes in river flow or direction, subsidence or erosion all cause rapid wetland degradation.

Coastlines are rarely static for significant periods of time and throughout the last 2 million years there have been major fluctuations of sea level. The latest rise, while certainly exacerbated by human activities which have contributed to the greenhouse effect, is actually a part of a long-established postglacial (or interglacial) trend. Unless coastal wetland ecosystems can adjust to the rate of change of sea levels, such as through increased sedimentation, they will be subjected to a major alteration of their water cycles, which may result in longer periods of inundation, saltwater stress, exposure to different, and potentially damaging, animal species and increased erosion. If the coastal topography is suitable the wetland fringe may simply retreat, but if grades are too steep, or if artificial flood defenses are put in place, the result will be progressive loss of salt marshes, mud flat complexes and mangroves.

Changing climate

Wetlands are highly sensitive to climatic change. This is particularly true of the world's peatlands, which cover large tracts of the Northern Hemisphere. The blanket bog and raised bogs of north-west Europe are particularly vulnerable to amount and frequency of rainfall. Any reduction in the precipitation/evaporation ratio is likely to lead to increased decay and eventual wastage of the peat mass. In the case of the vast peatlands of the tundra and taiga, a drop in the water table can result in increased carbon release, so contributing further to global warming.

15

Why we need wetlands

For 6,000 years, river valleys and their associated flood-plains have served as centers of human population, with many boasting sophisticated urban cultures. Their fertile soils brought in huge harvests upon which the peoples of the regions could depend. Today, the wetlands which nurtured the great civilizations of Mesopotamia and Egypt, and of the Niger, Indus and Mekong valleys, continue to be essential to the health, welfare and safety of millions of people who live by them.

All wetlands are made up of a mixture of soils, water, plants and animals. The biological interactions between these elements allow wetlands to perform certain functions and generate healthy wildlife, fisheries and forest resources. The combination of these functions and products, together with the value placed upon biological diversity and the cultural values of certain wetlands, makes these ecosystems invaluable to people all over the world. Within this section, wetland values are introduced in general terms. The Atlas section looks at specific wetland uses, illustrating the immensely varied wetland resources and values upon which people depend.

Flood control, water purification and shoreline stabi-lization, for example, can all be provided by wetland systems, while these simultaneously support fisheries and agriculture. However, most wetland development projects today concentrate on converting the ecosystem to maximise economic return from one sector such as agriculture or hydro-power. The limitations of this approach are now increasingly obvious. If wetlands are converted or developed without first taking into consider-ation their full value, the negative consequences can be felt immediately by local people. The economy of a region, or indeed a nation, may be affected adversely if the alterations are many or large. Development often requires major investments of capital, manpower, tech-nology, and inputs such as fertilizer, as well as substantial annual investments in maintenance. And where conversion is attempted, the ability of natural wetlands to sustain alternative development is found to be low; moreover, such conversions require a far more sophisticated system of management than is generally available to the average rural community.

Despite the importance of the range of resources and services which wetlands provide, we have tended to take these for granted. As a result, the maintenance of natural wetlands has received low priority in most countries. But even as apathy and ignorance continue to permit

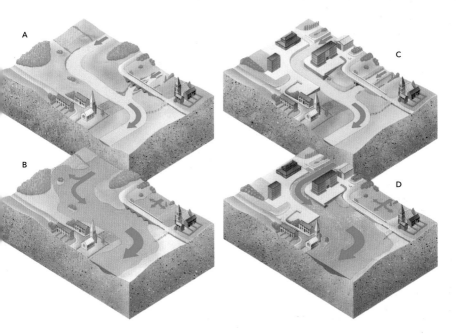

▲ *If floodplains are preserved [A], when floods occur [B], the floodwaters are free to spread out over the plain, causing little damage downstream. If, however, floodplains are developed [C], the floodwaters are channeled downstream [D], where the risk of damage is far more severe – caused not only by the flood heights, but also by the velocity of the floodwater as it flows downstream.*

conversion of wetlands, people are becoming increasingly aware of the loss of the services wetlands once provided free of charge.

Groundwater and flood control

When water moves from a wetland down into an underground aquifer (a rock deposit that contains water), it is said to recharge groundwater. By the time it reaches the aquifer, the water is usually cleaner, due to filtering processes, than it was on the surface. Recharge is also beneficial for flood storage because runoff is temporarily stored underground rather than moving swiftly downstream and overflowing.

Once in the aquifer, water may be drawn out for human consumption through wells, or it may flow laterally underground until it rises to the surface in another wetland as groundwater discharge. In this way, recharge in one wetland is linked to discharge in another. Wetlands that receive most of their water from groundwater discharge usually support more stable biological communities, because water temperatures and levels do not fluctuate as much as in wetlands that are dependent upon surface flow.

17

By storing rain and melted snow and releasing runoff evenly, wetlands can diminish the destructive onslaught of floods downstream. Preserving natural storage can avoid the costly construction of dams and reservoirs. In the Charles River of Massachusetts in the United States preservation of 15 square miles (38 square kilometers) of mainstream wetlands provides natural valley storage of flood waters. It is estimated that had 40 percent of these wetlands been reclaimed, the increased flood damage would have cost US$3 million each year. And had they been filled completely, the added flood damage would have been over US$17 million per year.

Stable shores and storm protection

Wetland vegetation can stabilize shorelines by reducing the energy of waves, currents or other erosive forces. At the same time, the roots of wetland plants hold the bottom sediment in place, preventing the erosion of valuable agricultural and residential land, and property damage. In some cases, wetlandsmay actually help to build up land. To illustrate the effect of wetlands in protecting against destabilizing sea forces, it was estimated in the United Kingdom in 1981, that sea walls

▼ *Young mangrove trees planted as part of a shoreline reclamation project in Jupiter Lighthouse Park, Florida, USA. The tree roots will help stabilize the mud and encourage sediment to accumulate.*

constructed behind salt marshes would be over 20 times cheaper to build than walls unprotected by salt marshes.

Hurricanes and other coastal storms cause wind damage and flooding. In the developed world, the principal consequence is property damage; in poor tropical nations, however, it is more often human injury and death. In Bangladesh, 150,000–300,000 people were killed in one storm surge in 1970. Many wetlands, in particular mangroves and other forested coastal wetlands, help dissipate the force and lessen the impact of coastal storms. The mangrove forest of the Sundarbans in India and Bangladesh, for example, breaks storm waves which often exceed 12 feet (4 meters) in height. In recognition of their protective functions, over the last 10 years the Bangladesh government has planted vast areas of mangroves to protect embankments and farms. After the Asian tsunami in December 2004 the prime minister of Malaysia called for protection of the country's mangroves because of their role in protecting coastal areas from the most severe effects of these tidal waves.

Sediments and nutrients
Sediment is often the major water pollutant in many river systems. Many wetlands help reduce this by serving as pools where sediment can settle. In addition, if reeds and grasses are present to slow down a river's flow, the opportunity for settling is increased.

Although the buildup of too much sediment in a wetland may alter its biological functions, such as floodwater storage and groundwater exchange, the quality of ecosystems downstream can be improved if suspended sediment is retained in the tributaries upstream. Because toxic substances, such as pesticides, often adhere to suspended sediment, they too may be retained with the sediment. Retaining sediment in upstream wetlands can lengthen the lifespan of downstream reservoirs and channels, and reduce the need for costly removal of accumulated sediment from dams, locks, powerplants and other man-made structures.

Nutrient retention and export
Wetlands retain nutrients, most importantly nitrogen and phosphorus, by accumulation in the subsoil, or storage in the vegetation itself. Wetlands that remove nutrients improve water quality and help prevent eutrophication – a process whereby the buildup of excess nutrients leads to rapid plant growth, increased oxygen demand and

ultimately reduced productivity and biological diversity. By removing nutrients wetlands can dispense with the need to build water treatment facilities. Under certain circumstances, wetlands can even be used for treatment of domestic waste from small, nonindustrial communities.

When wetlands remove nutrients (and pollutants) they are referred to as "sinks". This is particularly important with regard to nitrates which can be converted back to harmless nitrogen gas and returned to the atmosphere. When nutrients are returned to the surroundings, wetlands are said to act as "sources". A common role of wetlands during the growing season is to accumulate nutrients when the water flows slowly. These nutrients support fish and shrimps, as well as the forest, wildlife and agricultural wetland products. When water flows fast, wetlands may act as a source.

This cycle has important implications for algal growth, water quality, fish production and recreation downstream from the wetland area because it reduces nutrient levels at a season of the year when added nutrients are likely to cause eutrophic conditions downstream. Release of the nutrients occurs when they are less likely to cause eutrophication.

Many wetlands support dense populations of fish, cattle or wildlife which feed on the nutrient-rich waters or graze on the lush pastures. But in addition to this production

▼ *Indonesian farmer tilling ricefield with oxen. Many of Indonesia's tidal swamps have been converted to ricefields.*

within the wetlands, environments downstream and in coastal waters also benefit from the nutrients that are carried away by surface flow, in streams, or by ground water recharge.

In temperate regions, nutrients stored in growing wetland plants are released when the water cools and the plants die in winter. Part of the value of river and coastal fisheries is attributable to this vital support function provided by wetlands, and is an addition to their role as breeding or nursery areas for fish. Interfering with wetlands can disrupt nutrient production. After construction of the Aswan High Dam, for example, on the River Nile, export of nutrients and sediments was substantially reduced. As a result, total fish catch from the Mediterranean Sea adjacent to the Nile Delta decreased from 38,000 tons in 1962 to 14,000 tons in 1968. Fish catches have since increased however. This is believed to be due to increased nutrient inflows from agriculture and urban run-off.

▲ *Graceful progress across a Burmese lake. Wetlands as transportation systems can be overlooked by city dwellers. In the rural districts of many developing countries, water transportation often is the cheapest and most reliable method of carrying people and goods. In areas of permanent wetland, boats may be the only practical means of transportation.*

Further uses and resources

If our harvesting of wetland plants and animals respects the annual production rates and regenerative capacity of each species, we can enjoy the benefits of wetland productivity without destroying an important habitat. For example, direct harvest of the forest resources of many wetlands yields a number of products, ranging from fuelwood, timber and bark, to resins and medicines which are common nonwood, forest products.

Wetlands that contain important grasslands grazed by livestock are important to local communities. The Brazilian Pantanal, for example, supports over 5 million cattle. And over much of Africa, wetland grasslands provide critical dry-season grazing. Leaves, grasses, and seed pods may also be collected as fodder for sale, or used as a dry-season cattle feed.

Two-thirds of the fish we eat depend upon wetlands at some stage in their life cycle. In Africa, fish are the most important source of animal protein, making up 20 percent or more of the diet in many countries. Many of the fish are caught in small ponds and marshes on a daily basis.

In other wetlands, such as the Inner Niger Delta in Mali, there are annual fishing festivals toward the end of the dry season.

Wetland wildlife also provides an important source of protein and income even in the industrialized world. In Canada, the value of the mink, beaver and muskrat harvest exceeded US$43 million in 1976 and in many communities across North America this wildlife harvest continues to play an important economic and social role. Further south in Latin America, large numbers of water-birds, caimans and numerous other wetland wildlife are also harvested. For example in El Salvador, 30,000 tree duck eggs alone were harvested at Laguna Tocotal in one five year period, providing an important protein resource for the local community.

While many wetlands have been converted to intensive agriculture, many others, however, continue to be culti-vated in their natural form. Properly managed, natural wetland agriculture can often yield substantial benefits to rural communities. Indonesian tidal swamps have been cultivated successfully by communities using tradi-tional techniques. Rice is the major crop, often in combi-nation with coconuts and fruit trees, which help reduce

◄ Lake food chains begin with algae (1), prey to plankton like *Daphnia* (2), in turn prey to carnivorous plankton like cyclops (3). Minnows (4) reduce the chance of individuals being preyed upon by swimming in shoals. Perch (5), normally pelagic residents, breed in lit-toral waters, trailing mucus-covered eggs (6). Carp (7), however, are benthic, using whiskerlike barbels (8) to find food, such as caddis fly larvae (9), which build armor from plant fragments. Midge larvae (10) are prey for many ducks, fish and birds, particularly when they hatch into vast insect swarms. Tubifex worms (11) have oxygen-carry-ing haemoglobin in their blood and are a rich food source for fish. Underwater plants such as wormwort (12) and Canadian pondweed (13) are both bot-tom-rooted and photosynthesize in minimal light. Slime on their surfaces is food for the great pond snail (14). The great diving beetle (15) attacks any-thing from worms to tadpoles (16). Dragonflies (17) eat smaller insects; liv-ing underwater as carnivorous nymphs, they reach a last molt stage (18), before emerging as adults. Some lakeland birds, such as the great crested grebe (19) and tufted duck (20), feed by diving for fish, others such as the mallard (21) dabble for plant matter. The osprey (22) catches fish in its talons. Plants such as cattails (23) grow near the water's edge and help the slow colonization of the lake by the land. Long stems of lilies (24) convey oxygen to roots at the bot-tom. Floating plants, like the water sol-dier (25), have roots taking nutrients directly from the water. A water shrew (26) dives for beetles.

soil acidity. Corn, cassava and vegetables are also grown. In addition to these crops, swamp farmers also raise livestock (principally poultry), and maintain fisheries in the canals and ditches of the coconut gardens.

Energy resource

Some wetlands contain potential energy for human consumption, normally in the form of plant matter and peat. When this can be used on a sustainable basis, it is an important component of an integrated management scheme for the wetland. However, when extraction of peat is carried out on a large scale, there is a real danger of it destroying the ecosystem. Proposals for peat mining in Jamaica, Brazil and Indonesia, for example, have aroused considerable international concern in recent decades.

Biological Diversity

Many wetlands support spectacular concentrations of wildlife. In West Africa, floodplains in the Senegal, Niger and Chad basins support over a million waterfowl, many of them migratory, visiting the region during the European winter. And in Mauritania the tidal flats of the Banc d'Arguin National Park provide a wintering site for some 3 million shorebirds each winter. In Zambia, 30,000 black lechwe antelope (*Kobus leche smithemani*) inhabit the Bangweulu Basin, along with one of Africa's most important populations of sitatunga (*Tragelaphus spekei*)

◄ Green-backed heron (Butorides striatus) in Florida, USA. The wide variety of wetlands has led to the evolution of many specialized species. Second only to rain forests, wetlands contain a greater diversity of wildlife than any other terrestrial habitat, and a higher proportion of endangered species.

and shoebill storks (*Balaeniceps rex*). In Brazil, the Pantanal covers over 39,000 square miles (100,000 square kilometers), with large populations of spectacled caiman (*Caiman crocodilus*), capybara (*Hydrochoerus hydrochaeris*) and jaguar (*Panthera onca*), as well as one of the most distinctive mosaics of vegetation in Latin America.

While it is concentrations of individual species, rather than the diversity of species that has attracted most attention from conservation scientists, many wetlands do support a significant diversity of vertebrates, many of which are endemic (unique to the area) or endangered. The diversity of fish species is notable. For example, in East Africa, Lakes Victoria, Tanganyika and Malawi support over 700 species of endemic fish. In Lake Tanganyika alone, 214 species have been identified, 80 percent of which are endemic.

In many countries the inaccessibility of wetlands has attracted species which, although not confined to wetlands, are dependent upon the shelter they provide. In India and Bangladesh, the mangrove forest of the Sundarbans is the largest remaining habitat of the Bengal tiger (*Panthera tigris*). Over much of Latin America, wetlands such as the Pantanal of Brazil, Paraguay and Bolivia, and the Mesquitio lowlands of Nicaragua and Honduras provide the most important habitat for the jaguar.

Wetlands are also important as a genetic "reservoir" for certain species of plant. Rice, a common wetland plant, is the staple diet of over half of the world's people. Wild rice in wetlands continues to be an important source of new genetic material used in developing disease resistance and other desirable traits.

▲ Cutting peat from a bog on North Uist, Scotland. These vast peatlands represent a virtually inexhaustible supply of fuel for scattered communities. Small-scale use does not harm the environment, but in other places peat is mined extensively and put to industrial uses, resulting in extensive wetland loss and adding to greenhouse gases.

Adapting to life in wetlands

Many wetlands support lush habitats that provide countless species of plants and animals with a rich environment from which they can obtain most of their requirements. However, as strange as it may sound, the relationship between wetlands and water is not always simple and uniform. Plants and animals living there must adapt in order to survive alternate periods of flooding and desiccation, and the consequences this lack of uniformity brings.

Plant adaptation

Plants have developed a wide range of adaptations in order to survive and exploit their wetland environments. The most obvious developments are structural, concerned especially with the problem of supplying oxygen to roots growing in anaerobic (oxygen-deficient) soils or sediments. Most aquatic plants, such as water lilies (Pontederiacceae family), are extremely porous and contain special tissue, called *aerenchyma*, which has large, air-filled intercellular spaces. Oxygen diffuses 10,000 times faster in air compared to water, and so the aerenchyma are thought to facilitate the movement of oxygen from the leaves to the rootlike rhizomes.

Mangrove "knees"

Some trees, notably mangroves, such as *Avicennia* species, and the swamp cypress (*Taxodium distichum*), have evolved curious projections called *pneumatophores*,

◄ Bald Cypress trees in Bayou Cocodrie National Wildlife Refuge, Louisiana, USA. Bald Cypress has adapted to its habitat by evolving huge "knee" structures which stand above the water. Although these probably help provide oxygen to the roots, they are also thought to add stability as the trees grow to 160 ft (50 m).

▲ A brackish water area at Weeks Bay National Estuarine Research Reserve, Alabama, USA, showing a mix of salt and freshwater marsh species. Air passages i n the stems supply oxygen to the submerged roots. This adaptation was probably the first terrestrial plants evolved under similarly damp conditions about 400 million years ago.

or in the case of the swamp cypress, woody "knees". They develop from lateral roots growing near to the surface, and protrude vertically up to 12 inches (30 centimeters) above the soil or sediment surface. The precise function of these organs has been the subject of considerable scientific debate, but there is general agreement that they assist the plant in maintaining adequate root respiration. The root structures often produce dense networks which may also help to accumulate and stabilize sediment, providing a firm base on which the tree can grow.

Rather less prominent, but no less important, are the enlarged pores, called *lenticels*, which typify mangroves such as *Pelliciera* species. They are found on the bark and allow gaseous exchange between the atmosphere and the mangrove's tissue in the more or less continuously waterlogged environment of coastal swamps.

Plant chemistry
The external and internal adaptations which facilitate the movement of gases between plants and the atmosphere are important to only some wetland species. Others have

▲ Walter's Sedge
(Carex striata). The
4,000 species of sedge
all favor damp marshy
ground, and will
flourish even if their
roots are entirely
submerged.

developed biochemical adaptations, which scientists now think may be of much greater significance in terms of tolerating flood or waterlogged conditions. These complex biochemical adaptations work in one of two ways. Some species oxidize toxic elements, such as ions of iron (in the form Fe^{2+}), which diffuse into the root from the soil, altering their chemical structure and rendering them harmless. This oxidation process is thought to be highly efficient on the large internal and well-aerated surface of the aerenchyma tissue.

Other species of plants have developed mechanisms that excrete the toxic products of anaerobic respiration. Some, for example, diffuse acetaldehyde and alcohol, the toxic by-products, through the large surface area provided by finely divided roots. Other species, such as willow (Salix spp.) are capable of immobilizing or converting the harmful by-products to less toxic forms, such as, in the willow's case, pyruvic and glycolic acid.

Emergent plants

Rushes, reeds, sedges and some grasses typify wetland plants that are firmly rooted in the soil or sediment, but have erect stems that grow out of the water. Thus, while the stem is exposed to the atmosphere in the same way as any fully terrestrial plant, the roots have to cope with anaerobic conditions and, in some cases, salty environments. As with most wetland species air spaces in roots and stems allow oxygen to travel from the aerial parts of the plant to the roots. While land plants may have a porosity of about 2–7 percent of volume, up to 60 percent of many wetland plants is pore space.

Salt marsh species and mangroves also face the problem of salts. Some mangroves, including Rhizophora species, have specialized cells in the roots which block sodium while allowing essential elements, such as potassium, to move freely about the plant. Other species do not block salt at the roots, but have mechanisms for secretion. This is well illustrated in salt marsh grasses, such as Spartina species and some mangroves, the leaves of which are often covered with crystalline salt particles secreted from specialized glands.

Fuel efficiency

For the vast majority of plants, the first step of carbon fixation – part of the complex chemical process of photosynthesis in which carbon dioxide is converted into carbohydrates – is the production of a compound called phosphoglyceric acid. This is a three-carbon-atom compound, hence these plants are known as "C_3" plants. Many wetland plants, however, such as *Spartina* and *Panicum* species produce a four-carbon-atom compound called oxaloacetic acid.

Scientists have discovered that so-called C_4 plants have a much lower photorespiration rate than C_3 plants. And because photorespiration (light-stimulated respiration) is essentially a wasteful process (under some conditions, 30 percent of carbon reduced during photosynthesis can be reoxidized to carbon dioxide during photorespiration) C_4 plants are much more efficient than C_3 plants both in rate of carbon fixation and in water use. One of the benefits of efficient water utilization is that it reduces the rate at which soil toxins, present in generally high concentration in the oxygen-deficient wetlands, are drawn into the root network. This increases the opportunity for detoxification to occur in the thin oxygenated zone around the roots.

▼ *The edible cranberry* (Vaccinium macrocarpon) *grows in* Sphagnum *bogs and swamps. This species is heavily cultivated in the US and is used in a wide range of foods and health remedies.*

Successful reproduction

Some plants, such as roses, reproduce sexually. Others, such as strawberries, reproduce vegetatively, sending out "runners" which take root and form a "daughter" plant. Papyrus (*Cyperus papyrus*) can spread by sexual as well as vegetative means and is one of the fastest growing plants on Earth. The numerous spiky flower heads of papyrus produce thousands of seeds which fall into the water. Dispersal is controlled by the speed and extent of the water movment; in rivers this can be fast, while in isolated wetlands movement can be very limited and here dense patches of papyrus can develop. While the papyrus' sexual reproductive method colonizes new areas, its capacity for vegetative reproduction ensures older communities remain stable. New shoots are sent up from the rhizomes at regular intervals. Within the first three months they have grown, matured and died – but as one shoot dies the nutrients are withdrawn and passed on to the next generation in a continuous process of development. The efficient conservation of nutrients and its extraordinary rate of growth contributes to the plant's great success in many African swamps.

Peat-formers

The bog mosses (*Sphagnum* spp.) – also known as peat mosses – are exceptional wetland plants. They are capable of growing in very acidic conditions where, because there is no groundwater flow, the only supply of nutrients is from rainfall. Their structure enables them to hold enormous quantities of surface water, sometimes exceeding up to 15 times the weight of the plant itself. In this way a layer of *Sphagnum* moss can maintain high-levels of ground saturation even during prolonged dry periods. As the plant continues to grow upward, the lower parts die and become progressively compressed under the weight of accumulating vegetation. The slow rate of decomposition, thanks to the acidic conditions, makes *Sphagnum* an exceptional peat-former and maintainer. These species are particularly characteristic of the raised bogs, blanket mires and muskeg terrains of the oceanic, high-altitude and high-latitude regions.

Surface floaters

Many wetland plants are surface-floating species. Apart from the well-known water lilies (*Nymphaea* and *Nuphar* spp.), these include the water lettuce (*Pistia stratiotes*), water fern (*Azolla nilotica*), numerous members of the duckweed family (Lemnaceae) and water hyacinth (*Eichhornia crassipes*). Some, like the water hyacinth, have high growth rates. A native of South America, the introduction of water hyacinth to other subtropical and tropical regions, including the southern United States (where it is known as the "Florida devil"), India ("Bengal terror"),

▼ *Water-hyacinth (Eichhornia crassipes) in bloom. This plant thrives in wetlands where pollution has increased nutrient levels. If left unchecked, the fast-growing plant can engulf a water body, blocking sunlight and oxygen and slowing the water's flow. The consequences can be devastating for the resident wildlife.*

Africa and Southeast Asia, has enabled this adaptive plant to exploit new territory, causing widespread problems of weed control.

Water hyacinth grows rapidly into extensive surface mats especially in sheltered inlets or oxbows. Floods, wind and wave action can break up the mat, causing it to spread rapidly in rivers and lakes. Floating mats can become so large that the force of the vegetation is sufficient to damage man-made structures. Navigation is sometimes completely impossible and drainage may be impeded. Where water hyacinth has hampered drainage in parts of Africa, habitat conditions are often created that are ideal for carriers of disease such as bilharzia-carrying snails and malaria-transmitting mosquitoes.

Although the waters infested by excessive growth of water hyacinth becomedepleted in oxygen, there can be some benefits. The plant is a highly effective absorbent of nutrients, such as nitrogen and phosphorus, together with contaminants and other toxic wastes. In Southeast Asia, the water hyacinth is used as food for cattle and other livestock.

Animal adaptation
Many animals, like plants, have adapted well to an aquatic existence. Their adaptations have naturally centered

▲ *Gray-white nymphs of Taosa plant hoppers feed on heavily damaged water-hyacinth. In the Amazon such insects help to regulate the mat's expansion. The sap-sucking insects not only damage and weaken the weed, they introduce plant pathogens.*

around breeding and feeding in a wetland environment. To exploit the many rich wetland habitats, such as estuaries and mud flats, animals have evolved their own unique adaptations to ensure that they can exploit one particular niche more effectively than their competitors.

Insect adaptations

To survive in wetlands, insects have developed a host of adaptations. Some, such as dragonflies and damselflies, for example, lay their eggs in the tissues of submerged plants. Water boatmen (family Corixidae) and back-swimmers (family Notonectidae) have developed strong legs fringed with hairs that propel them through the water as oars do a boat. Many aquatic insects, including mosquitoes, when at their larval stage, have tiny tubes attached to the abdomen that poke out of the water, allowing the larvae to breathe. Others, such as the great diving beetle (*Dytiscus marginalis*), breathe by trapping air bubbles under their wings before each hunting dive. The larvae of some species, such as caddisflies (*Limnephilus* spp.), obtain oxygen from the water itself. They absorb the oxygen through thin sections of their fine outer body casing, known as the cuticle.

Water beetles and water boatmen are highly efficient hunters, pursuing their prey through the underwater jungle with speed, grace and agility. Other insects have adapted to life on the surface film, which provides a rich hunting ground for many land animals falling into the water. The pond skater (*Gerris-lacustris*) has specialized legs that enable the insect to literally skate over the surface of the water. The limbs' lower segments are wide and paddle-shaped, with hairs that repel water and increase the surface area. Whirligig beetles (family Gyrinidae) also swim on the surface, but unlike the pond skater, they are half submerged. Their eyes have compensated for this half-in, half-out arrangement by being divided into two, allowing them to see both above and below the surface of the water. Both groups of insects can quickly cover large

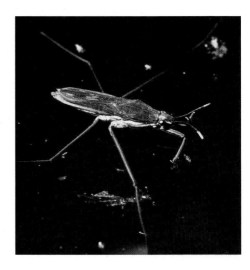

▼ *Pond skater resting on the surface of a pond. Adaptations to the ends of the skaters' legs enable them to stand on the water surface without breaking the "skin" formed by surface tension. Standing on the surface enables the skaters to detect ripples produced by heavier and less well-adapted prey struggling in the water. By homing in on the source of the ripples, the skaters are assured of a meal.*

areas of stillwater, scavenging the unfortunate animals unable to escape the force of surface tension.

Crustaceans and carnivores

Among invertebrates, it is the crustaceans that show some of the most sophisticated adaptations to aquatic life. The water fleas *Daphnia* are highly mobile free-living crustaceans that inhabit shallow ponds and ditches. They have developed appendages specially adapted for filter-feeding. Close-set rows of long, fine, feathery bristles cover the limbs attached to the thorax (the middle section of an insect). The bristles remove coarse particles from detritus or plankton and pass them backward to be either swept away with the water current or broken up for food by the limbs further back along the body. The gathered material is passed toward the mouth via a "food groove" which runs along the flea's underside.

A rich variety of carnivores exists even among the invertebrates. As well as the many diving beetles, pond skaters and water bugs, there is one particular group, the dragonflies, whose nymphs have evolved a highly efficient weapon. They lie concealed in vegetation and seize passing prey with a "mask" that shoots out on a hinged arm and grabs the victim between two sharp hooks.

▲ *Although mosquito larvae (Culex pipiens) have an aquatic life style, they have no specialized organs for breathing underwater. In order to take in oxygen, most species pierce the surface with a thin tube that grows from the abdomen. The "grip" of surface tension is sufficient to hold the animals at the surface, so that they can maneuver to take in food particles while breathing.*

33

Amphibians

Water is essential for amphibians. Although they can spend much of their time on land, amphibians, because their eggs are unprotected against drying out, must return to the water to breed. Tailed amphibians include the newts and salamanders. The largest European newt is the crested newt (*Triturus cristatus*), which is found in slow-moving waters with dense aquatic vegetation. The female lays between 200 and 400 eggs singly on the submerged leaves of grasses or aquatic plants. The eggs hatch after two to three weeks. The larvae (tadpoles) have external gills and live in water for some three months before metamorphosis is completed. The adults live on dry land outside the breeding season, but utilize the cover of stones, wood or moss and feed on worms, snails and insect larvae. Young newts may not mature for two to three years and do not return to the water to breed before then.

Frogs and toads make up the tailless amphibians. They are found in wetlands throughout the world, and occupy a wide range of altitudes, water salinities and qualities, as well as vegetation types. Some, such as the edible and marsh frogs (*Rana esculenta* and *R. ridibunda*), are found mainly in low-lying wetlands; the common frog (*R. temporaria*) lives in mountainous regions of Europe and

▲ *A tree frog* (Hyla arborea) *uses the specially adapted pads on its hands and fingers to cling to a plant stem. Amphibians are ideally suited to wetlands. While some remain in the water all their lives, others return only to breed.*

Asia. The moor frog (*R. arvalis*) thrives in peat and swamp conditions, and ranges from northwest France to Siberia as far as the Arctic Circle.

Tree frogs (family Hylidae) such as the common tree frog (*Hyla arborea*) live in lush vegetation in most tropical and subtropical regions of the world except Africa. These frogs have specialized hands and fingers that have circular pads from which is secreted a gluey substance. This helps them cling to the slippery leaves of trees and shrubs, where they capture insects often by means of a distinct leaping action. Tree frogs worldwide are typically small and vivid green in color. Their coloration acts as a camouflage, matching the foliage with which they are normally associated.

Reptiles

Reptiles such as snakes, iguanas and crocodiles are common in the world's subtropical and tropical wetlands. Some, such as the grass snake (*Natrix natrix*), also live in the high-latitude wetlands of the Northern Hemisphere. Other species of *Natrix*, such as the Dice snake (*N. tessellata*), are well adapted to wetlands, being extremely proficient swimmers and particularly adept at catching fish. Active by day as well as night, the snake is found in freshwater and wetland edges from Europe to Asia. Like

▶ *Despite its name, the grass snake (*Natrix natrix*) has decidedly aquatic habits, and is frequently found in damp and marshy areas. The snake is an excellent swimmer, and frogs constitute a major part of its diet. Its European relative (*Natrix viperinus*) is even more aquatic and feeds almost exclusively on fishes. The North American moccasins (*Natrix fasciatus*) have a similar semiaquatic life style, as do the North American garter snakes (*Thamnophis sp.*)*

the grass snake, however, it needs to lay its eggs on land and so is dependent on a habitat where land lies close to its watery hunting ground.

Some wetland snakes, such as the cottonmouth moccasin (*Agkistrodon piscivorus*) with its distinctive white inner mouth, are highly poisonous. This particularly dangerous species of snake inhabits the cypress swamps of the southern United States and feeds primarily on frogs, small fish, salamanders and crayfish.

Crocodiles and alligators

These are the world's largest reptiles and are a key feature of wetlands from the warm subtropical regions of the Americas to the tropics of Africa, Southeast Asia and Australia. Unlike most animals, crocodilians continue to grow throughout their lives – some males grow to over 13 feet (4.5 meters) and weigh over 500 pounds (225 kilograms). It is thought that crocodilians can live to be over 100 years old and for this reason they represent a force for considerable stability within the wetland wildlife community.

Crocodilians are effective hunters. They can move surprisingly quickly onland, and the whipping motion of their long, muscular tails makes them efficient swimmers. Bony plates (osteoderms) provide an armored shield

◀ *An American crocodile (Crocodylus acutus) at rest. The crocodile can rest with its mouth flooded – a muscular valve keeps water out of the throat. As long as the valve remains closed, the crocodile can breathe freely through its nostrils. The American crocodile is less aggressive than both the Nile and Australian crocodile and is rarely seen in the wild.*

▼ *Beaver dam. Beavers build "lodges" of trees and branches above water level and dam streams and rivers with stones, sticks and mud.*

which protects them from virtually all predators. With eyes and nostrils located on top of their heads, crocodilians can breathe and look around with only a small part of their body protruding from the water. This enables them to approach shoreline animals without being seen. Once the jaws take hold of the prey, release is virtually impossible and the animal is then usually dragged under water and drowned. The muscles which close the jaw are extremely strong and the array of teeth can be replaced when worn out or lost. A single animal may go through 3,000 teeth in its lifetime.

Hard times
The water cycle is as critically important to alligators and crocodiles as to all animals of the wetland habitat. Alligators in southern Florida excavate the marl (a fine-grained clay or loam rock) and soft, weathered limestone bedrock to create depressions commonly called "gator holes". The deeper water is particularly important in the dry season when fish and other aquatic organisms concentrate in these refuges. This provides the alligator with a readily available food supply and is attractive also to a wide range of fish-eating birds, such as heron and osprey.

During the breeding season, different species build different types of nests, but all are designed to keep the eggs warm until they hatch. Most crocodiles dig a hole in suitable ground and deposit the eggs in layers which are carefully covered with sand. The nest needs to be close to water to enable the female to splash water on to it to keep it cool.

Alligators build their nests above ground from leaves, branches and sediment. The material is shaped into mounds up to 6 feet (2 meters) in diameter and 3 feet (a meter) high. Eggs are laid in the center and as the organic material decomposes sufficient heat is generated to incubate them.

Small aquatic mammals
There are a number of small mammals that are specialized for life in and around water, and that show both physical and behavioral adaptations to their environment. One of the best examples is the two species of beaver, the American beaver (*Castor canadensis*) and the Eurasian beaver (*C. fiber*). These large rodents are excellent swimmers;

to help propulsion they have webbed rear feet and a broad, flat tail, which they use like an oar. The thick fur of beavers provides both insulation and waterproofing. A beaver is also able to close its nose and ears when under water. The most notable behavioral adaptation of beavers is the building of dams to create the pools in which they build their lodges.

Otters, such as the Eurasian otter (*Lutra lutra*), have many of the same physical adaptations as beavers to life in water, such as webbed feet and waterproof fur. They have the added advantage of a slim body, which undulates easily to help swimming. Other small mammals that are well adapted to an aquatic life include the muskrat (*Ondatra zibethicus*), the fish-eating rat (*Ichthyomys stolzmanni*), the marsh mongoose (*Atilax paludinosu*) and the otter-civet (*Cynogale bennetti*).

Larger aquatic mammals

Perhaps the most famous large aquatic mammal is the hippopotamus (*Hippopotamus amphibius*). While the popular image of the hippo is one of an animal that seems to be permanently submerged up to its nose in water, this masks the hippo's vitally important role as a carrier of nutrients from land to water. Hippopotamuses graze on terrestrial vegetation, generally at night, but

▶ The hippopotamus has an almost completely aquatic life style and has become superbly adapted to life in rivers and other wetlands. The nostrils are positioned on the upper surface of the snout and protrude so that the hippo can float with almost all of its bulk below the surface.

◄ *North American river otters* (Lutra canadensis) *at play. River otters can be found living by streams, rivers, lakes, estuaries, and salt and freshwater marshes. They eat fish, crayfish, frogs, turtles and aquatic invertebrates, plus an occasional bird or rodent.*

defecate in the water, where they remain for most of the day to keep cool. Their tails propel and help disperse the nutrient-rich waste products into the water to the benefit of the aquatic animals.

Male hippos also spray dung to mark their land territories. Harvester termites (family Hodotermitidae) then collect particles of the dung to mark their own land territories. The termites collect the dung because it still contains partially digested grass fragments. The termites complete the breakdown of the grazed grasses and return the nutrients to the floodplain soils. In addition, fungus-growing termites (*Macrotermes* sp.) build mounds on the floodplains, and as the mounds are broken down by erosion they enrich the soil. This nutrient-rich soil is favored by the shorter, more nutritious grasses, which are the preferred food of the harvester termites. Thus, the recycling of nutrients, involving hippos, termites and grasses, gradually enriches the floodplain soils.

Fleet-footed herbivores

The sitatunga (*Tragelaphus spekei*) is an elusive, shy, well-camouflaged antelope. It is the only large mammal

to feed on and inhabit the papyrus swamp of central and East Africa. Living as far south as the Okavango Delta, it can move across the dense papyrus beds and soft terrain withsome ease despite its relatively large size – males weigh over 220 pounds (100 kilograms). Long hooves, almost twice the length of those of similar-sized antelopes, spread out as it walks, providing good mobility in the wetland habitat. The sitatunga is further favored by raised hindquarters, giving a slow, loping gait enabling it to move quietly around the thick reedbeds – an important advantage in avoiding predators. The sitatunga also avoids capture by submerging itself in water, with just its nostrils above the surface.

Sitatunga feed on grasses and other emergent wetland plants in addition to papyrus. This adaptability is important as annual movements as well as diet are determined by the flood pattern. When the water is high, the antelopes move toward the dryland margins to feed on a variety of vegetation; when water levels drop the sitatunga move deeper into the swamp to survive almost exclusively on papyrus shoots. Such versatility is a key to survival.

Sitatunga raise their young on platforms of flattened vegetation, hidden in the extensive areas of swamp.

▼ *Waterbuck (*Kobus ellipsiprymnus) *resting in East Africa. Although not as trully aquatic as the sitatunga or lechwe, where possible waterbucks graze only on aquatic plants.*

Normally, however, well-trampled paths form a network indicating favored feeding areas and routes. The main threat in the water is from crocodiles, but lions and leopards may attack on islands and the young taken by pythons.

Antelope dominate the seasonally flooded grasslands of East Africa. One of the most abundant is the red lechwe (*Kobus leche leche*), sometimes described as a semi-aquatic antelope because of its particular ability to graze on the young shoots of sedges and other plants growing in shallow waters. Like the sitatunga, the red lechwe have elongated hooves for moving over soft ground. In fact, they can run faster through shallow water than on dry land. Their preference for the flooded environment stems from a combination of the availability of green vegetation and the capacity to escape from land predators such as lions, leopards, wild dogs and humans.

Fish

Rivers, lakes and coastal waters are home for an enormous diversity of fish species, most of which are dependent in some way on wetlands – whether for food, spawning, nursery or other habitat requirements.

Some species, such as the European carp (*Cyprinus carpio*), for example, migrate from rivers to seasonally inundated floodplain forest or grassland for spawning; others migrate from deepwater lakes to vegetated shallows; yet more move from the open sea to coastal mud flats, lagoons, marshes and mangroves.

In the case of migratory species such as the salmon (*Salmo salar*), the distances covered may be thousands of kilometers from ocean feeding grounds to the shallow gravel-bed "redds" (spawning grounds) of streams. Both American and European eels (*Anguilla rostrata* and *A. anguilla*) spend most of their lives in fresh or brackish water but return to the Sargasso Sea to spawn.

Distinctive groups of fish are adapted for life in different water conditions. The fish population of rivers often reflects speed of flow. The fast, or torrent, zone has clear, well-oxygenated, mineral-poor water, which flows over generally high, rocky terrain. This zone is inhabited by fish such as trout (*Salmo trutta*) and salmon (*S. salar*). Medium waters are characterized in northwest Europe by barbel (*Barbus barbus*), chub (*Leuciscus cephalus*), roach (*Rutilus rutilus*), rudd (*Scardinius erythrophthalmus*) and perch (*Perca fluvialis*).

In the intermediate zone between medium and slow-moving waters, pike (*Esox lucius*), carp, tench (*Tinca*

tinca) and white bream (genus *Abramis*) may be more typical. Where there is little or no grade, the current is tidal and the water brackish to saline, flounder (*Platichthys flesus*) and mullet (for example *Crenimugil labrosus*) are usually widespread. They are able to tolerate seawater as well as freshwater during times of flood and at low tide. Migratory species such as salmon, trout and eel can survive all levels of salinity and are commonly described as *euryhaline*.

High and dry

Of the several species of lungfish, the African lungfish (*Protopterus aethiopicus*) is the largest. It grows to about 6.5 feet (2 meters) long and lives in the rivers and lakes of East and central Africa. Other species of lungfish are found in Australia and South America. As their name suggests, lungfish have lunglike breathing organs with which they can breathe air to supplement the usually poor oxygen supply they receive from their swampy river and lake habitats. In addition, these fish live in regions

▲ *The omnivorous Alaskan grizzly bear (Ursus arctos) has learned to take every advantage of its varied habitats. These bears have caught Pacific salmon that were making their way up the McNeil River to their spawning ground. The migration of the salmon and their subsequent death upriver provides an annual bonanza of protein for the bears.*

that experience seasonal periods of flooding and desiccation. When the water level becomes dangerously low, the lungfish digs a burrow, covers itself in a slimy mucus secretion to prevent its skin from drying out and breathes air. The fish will then slow its metabolic rate to conserve energy until the water rises again. This form of inactivity is known as *aestivation*.

Shorebirds, ducks and geese

"Shorebirds" (or "waders") is the name given to a group of 214 species of small to medium-sized long-legged birds that prefer shallow water and wetland habitats during the nonbreeding season, and are usually gregarious and often highly migratory. The majority of the 214 species of shorebird belong to one of two families – the Charadriidae, which is made up of the lapwings and plovers, and the Scolopacidae, which comprises the sandpipers and snipes. Particular adaptations have evolved within the shorebird group that enables the various species of bird to live and breed together successfully in their wetland environment.

Great cormorant
(*Phalacrocorax carbo*)

Expert divers The gulls and terns exploit the first top few centimeters of water, cormorants dive more than 40 meters (130 feet) to reach their prey.

Herring gull
(*Larus argentatus*)

Beaks and feet

Beaks of the shorebird group are highly adapted and sufficiently versatile to obtain food of various types in highly efficient ways, often exploiting particular food niches at different depths in mud flats. This avoids extreme competition for food and enables very high numbers of a variety of bird species to feed in the same physical space.

Curlews have a long, strong curved bill which can be probed deeply into the mud for food. Snipe do the same, but their beaks are long, straight and slender, enabling more sensitive deep-probing of soft sediment. The avocet is one of the few birds with an upturned bill which enables it to explore the upper layers of the mud.

Specialized feet are necessary for many birds to survive in their wetland environments. Greenshanks, redshanks and most other waders have long toes, spread wide apart to increase the

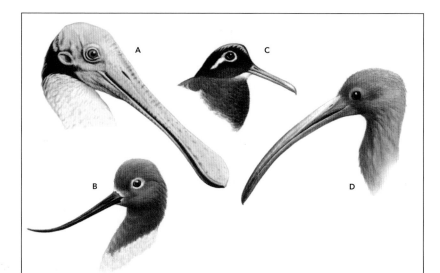

The beaks of waterbirds are adapted to their feeding habits. The roseate spoonbill [A] (*Ajaia ajaia*) rakes through bottom ooze to trap invertebrates. The American avocet [B] (*Recurvirostra americana*) has an upturned beak which it uses in a sideways motion to skim vegetation from muddy water. The painted snipe [C] (*Rostratula benghalensis*) uses its beak to pluck worms, insects and seeds from marshy ground and lake shores. The scarlet ibis [D] (*Eudocimus ruber*) probes deep into coastal mud flats and mangrove thickets in search of crabs and mollusks.

surface area of the foot. This allows the birds to move freely across soft mud and other sediment, where other birds and animals might become bogged down.

Other species of waterbird have developed legs for swimming and diving. Those of ducks and geese are near the rear of the body, where they can be used more effectively as paddles. The disadvantage of this arrangement for movement on land is only too apparent. The inconvenient distance between the legs and the bird's center of gravity results in the familiar ungainly "waddle". In the case of the osprey the scales on the footpads are armed with small spines which help the bird to secure and carry slippery fish.

Diving techniques
While gulls and gannets can use the momentum of a sky dive to overcome the resistance of water, birds that have

to dive from the surface must rely on the rear leg positioning, which is so marked in grebes and diving ducks, to give unimpeded thrust. Underwater swimmers are usually heavier than land birds of a similar size. Additional reduction in buoyancy can be obtained by compressing feathers or air sacs, forcing out the air. The dabchick uses this technique to float at different levels in the water. The dipper, however, can walk along the bottom of a stream by facing the flow, tilting up its back and allowing the force of water to hold it down. Its paddle-shaped toes and webbed feet help the dipper to walk through water.

The webbed toes of ducks, geese and divers present an enlarged surface area for powerful propulsion across the surface or underwater. Some sea-going ducks, such as scoters and other birds, including cormorants, use not only their legs but also their wings for swimming. Different species of diver tend to utilize different depths, so that like the waders, which have different bill lengths and structures, not all have the same underwater niche. While gulls and terns exploit the first top few centimeters of water, cormorants dive more than 130 feet (40 meters) to reach their prey.

▼ *The darter (Anhinga anhinga) fishes quiet inland waters. Darters swim with their bodies below the water and only their heads and necks above the surface.*

Wetlands loss

When Christopher Columbus set sail from southern Spain in August 1492, he epitomized the spirit of a continent on the threshold of the modern era. In the years that followed, mariners and explorers visited every corner of the globe searching for land and riches to serve the developing states of Europe. Simultaneously, the same creativity and enterprise in agriculture and industry drove rapid changes across the continent.

The agricultural and industrial revolutions not only changed the social, economic, and political face of Europe and ultimately the world, but they also set into motion a process of ecological change and devastation which continues today. In the words of one historian, the Columbian achievement "enabled humanity to achieve, ... the transformation of nature with unprecedented proficiency and thoroughness, ... altering the products and processes of the environment, modifying systems of soils and water and air". Among the systems that received special attention from this thriving European culture, both at home and in the newly discovered and settled

▼ *Houses built on salt marsh in Groton, Connecticut, USA. Suburbanization in the United States has*

lands, were the marshes, floodplains and other wetlands, most of which were perceived to be disease infested and obstacles to development. Once drained, these could be put to productive agricultural use.

Working with water

Europe during the 1400s was, of course, not the first civilization to modify natural wetlands. Presently, almost 70 percent of the world's population lives on sea coasts, and in many regions, river valleys and lake shores have been settled for thousands of years. Over much of Asia, a diversity of cultures and empires had been built upon the control and exploitation of the regions' wetland systems. The civilizations of the Indus Valley and Angkor in Indochina drew much of their economic strength and stability from their efficient manipulation of the Indus and Mekong rivers.

In studying the use of water in Asian society, historians have made clear distinctions between hydraulic civilizations, which controlled water flow through dikes and dams, and aquatic civilizations, which exploited the annual cycle of river flood and adjusted to its excesses by, for example, building their houses on stilts. In general, the hydraulic form of life developed inland, where water was more seasonal and needed to be controlled in order to be brought most efficiently to the best agricultural land. In contrast, the aquatic civilizations inhabited the deltas and floodplains where water was abundant as it moved to the sea.

The concept of working with nature was largely absent from Columbian Europe. Instead, the hydraulic culture which sought to control and dominate the aquatic environment governed much of human society's relationship with wetlands in the subsequent 500 years. Along the coast of Europe, the Dutch began to dike their shallow sea and turn its bed into reclaimed agricultural land, known as polders. In Britain and France, similar investments were made in channelling the major rivers. And as the new colonies became established in the Americas, Africa and Asia, this hydraulic technology was exported, and indeed continues today.

Agricultural devastation

The loss of wetlands worldwide, which some specialists estimate as being in the order of 50 percent of those that once existed, is the direct consequence of this hydraulic vision of the world. And although detailed figures are

placed additional pressure on wetlands that require relatively little effort to prepare them for housing.

▲ Forest killed by the effects of acid rain on Mt. Mitchell in western North Carolina, USA. Produced in the atmosphere from industrial pollution, acid rain has a double deadly effect. Falling as precipitation, the acid chemicals attack the exposed surfaces of vegetation, destroying foliage. Acid rain run-off is likely then to enter the groundwater, affecting the hydrology of many wetland habitats. In particular, it collects in ponds and lakes, killing off the flora and fauna.

difficult to obtain, the limited information tells a sorry tale. In the United States alone, some 52 percent – 360,000 square miles (870,000 square kilometers) – of the wetlands which once existed are believed to have been lost, while in some states the proportion lost is even higher. Taking the nation as a whole, 80 percent of wetland loss has been to agriculture. For 200 years, the conversion and destruction of wetlands was actively encouraged by the United States federal government. In 1763, George Washington set up a company to drain the Great Dismal Swamp of Virginia and North Carolina, and convert it to agricultural land. Although that particular scheme failed, the nation's attitude toward wetlands was set. Further legislation in the mid-1800s, known as the Swamp Land Acts, underlined the conversion attitude. States were encouraged to build levees, carry out drainage projects and destroy the mosquito-infested wetlands. Following these acts alone, 100,000 square miles (260,000 square kilometers) of wetlands were lost.

In Europe, the rates of loss are less well documented, but, with the continent's high population density and longer history of economic development, the conversion of natural ecosystems is believed to be greater. Forty per-cent of the coastal wetlands of Brittany have been lost since 1960, and two-thirds of the remainder are seriously affected by drainage and similar activities. In southwest France, some 80 percent of the marshes of the Landes have been drained. In Portugal, some 70 percent of the wetlands of the western Algarve, including 60 percent of estuarine habitats, have been converted for agricultural and industrial development. In New Zealand, it is estimated that over 90 percent of wetlands have been destroyed since European settlement, and drainage continues. Fourteen percent of the remaining freshwater wetlands were drained in the North Island in the five years between 1979 and 1983. For the developing world little detailed information is available on rates of wetland loss. However, what is available has given rise to concern that entire ecosystems are now under threat. In the Philippines, some 1,000 square miles (3,000 square kilometers) – 67 percent – of the country's mangrove resources were lost in the 60 years from 1920 to 1980, some 650 square miles (1,700 square kilometers) being converted to ponds for farming shrimp and milkfish. In Nigeria, the floodplain of the Hadejia River has been reduced by more than 115 square miles (300 square kilometers) as a result of dam construction; and in Brazil most estuarine wetlands have been degraded as a result of pollution.

The consequences of wetland loss

As we have seen, many wetland regions have been destroyed because society has viewed their destruction as either good in itself, or as a small price to pay for the benefits expected from wetland conversion. Today, such

▶ *Rubbish dumped in a wetland. The unoccupied appearance of many wetlands causes people to consider them "natural" dumping grounds for domestic waste, unwanted vehicles and construction materials. Once dumping starts, it is very difficult to stop people continuing to dump.*

policies are increasingly recognised as short-sighted, and both socially and economically indefensible.

Dams, of which there are 114 major projects planned in West Africa alone, and other river basin schemes have come under special criticism. Many have destroyed wetlands while falling far short of their predicted benefits. The result has been hardship for those populations dependent upon the floodplain and other wetlands downstream. In Nigeria, fish catches and floodplain harvest declined by over 50 percent in an area extending 120 miles (200 kilometers) downstream of Kainji Dam, and some 1,000 tons of yam production were lost in the lower Anambra Basin of eastern Nigeria. In northern Cameroon, construction of the Semry II irrigation project on the Logone River reduced flooding downstream, causing a collapse in the fish yields and making it impossible to grow floating rice, the principal cereal crop of the Kotoko community.

The issue of floodplain development highlights the general principle that wetlands will be destroyed where people envisage putting their water to a more productive use. But just how valid is this assumption? When efficiency is measured in terms of profit per unit of water, data from African

▼ *Glen Canyon Dam, Arizona, USA. The dam is typical of the many hydro-engineering projects that have been undertaken during the last 80 years. Such schemes usually have wholly admirable objectives, such as flood control or the provision of electricity. However, a large dam can radically alter the pattern of flooding in a river system, resulting in disastrous consequences for downstream wetlands.*

floodplains suggest that over time there is very little differ-ence between traditional extensive methods of agriculture combined with fishing and grazing, and intensive rice culti-vation. And when the costs of the capital investment are taken into account rice cultivation can actually lose money.

A question of economics
Some of the products and services of wetlands are sold, such as commercial fisheries, meat and skins from grazing herds and crops. But many wetland values do not have identifiable markets, such as water purification and flood protection. Because these values are "free goods" they tend to be ignored in the economic calculations that decide whether wetlands should be conserved or developed. The result unfortunately usually favors devel-opment and, with it, wetland degradation. Private landowners, for example, frequently decide to drain their wetlands because they expect to earn more from growing crops than from leaving them in their natural condition. They may be perfectly aware of the role wetlands play in groundwater recharge and discharge, flood control, fisheries support and nutrient retention, but these public benefits count for less than their private profit.

▲ *Sprinkler irrigation system in private agricultural land along the John Day River, Oregon, USA. Spraying is a wasteful method of irrigation as losses through evaporation are very high. In many regions, aquifers are being depleted for irrigation, and the hydrology of entire regions is threatened.*

CAUSES OF WETLAND LOSS							
HUMAN ACTIONS							
Direct	Estuaries	Open Coasts	Flood-plains	Freshwater marshes	Lakes	Peatlands	Swamp forest
Drainage for agriculture and forestry; mosquito control	●	●	●	●	○	●	●
Dredging and stream channelization for navigation; flood protection	●	—	—	○	—	—	—
Filling for solid waste disposal; roads; commercial, residential and industrial development	●	●	●	●	○	—	—
Conversion for aquaculture/mariculture	●	—	—	—	—	—	—
Construction of dikes, dams and levees; seawalls for flood control, water supply, irrigation and storm protection	●	○	○	○	○	—	—
Discharges of pesticides, herbicides and nutrients from domestic sewage; agricultural runoff; sediment	●	●	●	●	●	—	—
Mining of wetland soils for peat, coal, gravel, phosphate and other materials	○	○	○		●	●	●
Groundwater abstraction	—	—	○	●	—	—	—
Indirect							
Sediment diversion by dams, deep channels and other structures	●	●	●	●	—	—	—
Hydrological alterations by canals, roads and other structures	●	●	●	●	●	—	—
Subsidence due to extraction of groundwater, oil, gas and other minerals	●	○	●	●	—	—	—
NATURAL CAUSES							
	Estuaries	Open Coasts	Flood-plains	Freshwater marshes	Lakes	Peatlands	Swamp forest
Subsidence	○	○	—	—	○	○	○
Sea-level rise	●	●	—	—	—	—	●
Drought	●	●	●	●	○	○	○
Hurricanes and other storms	●	●	—	—	—	○	○
Erosion	●	●	○	—	—	○	—
Biotic effects	—	—	●	●	●	—	—

— Absent or exceptional

○ Present but not a major cause of loss

● Common and important cause of wetland degradation and loss

To compound the problem, many governments which on the surface actively encourage wetland conservation, have, by some of their policies, caused the opposite to happen. For example, in the countries of the European Union, all of which are parties to the Ramsar Convention, wetland drainage was stimulated by the artificially high prices paid for a number of crops under the Common

Agricultural Policy (CAP). While over time many of these subsidies have been removed, others remain and there is continued concern that the CAP, combined with regional economic integration, remains an important driver for wetland loss.

Irrigated land, much of which was formerly natural wetland, is also meeting serious problems in many areas of the world. It has been estimated that salinization, waterlogging and alkalization of soils affect some 50,000 square miles (1400,000 square kilometers) – 20 percent of the world's irrigation schemes. As a result, hundreds of thousands of square kilometers of irrigated land, and the revenue from the crops, are being lost each year. In other instances, irrigation was achieved at the cost of disrupting normal water supply and consequently has compromised the long-term viability of the investment. In Peninsular Malaysia, 90 percent of freshwater swamps have been drained for rice cultivation. However, without the freshwater normally supplied by the swamps, rice production has often been below expectations.

In the past, loss of these wetland benefits generally has gone unremarked because the relatively strong national and household economies of industrialized nations could afford to pay for the consequences. Flood

▼ *Boys collect illegally cut mangrove trees at the site of a future shrimp farm, Honduras. In many parts of the world, the financial benefits of converting wetlands to other uses has led to the loss of these wetlands and of the public environmental benefits they provided free of charge.*

control and water purification, once provided free by wetlands, are today maintained by dams, dikes and other measures that were financed by increased taxes.

The increasing cost of wetland loss to the industrialized world, however, has reached such proportions that major efforts are now being made to conserve the remaining wetlands as functional economic units. For example, in the United States, where the loss of privately owned wetlands has resulted in major public costs, federal aid for drainage activities has now been removed, and crop subsidies are no longer available for land-owners who drain their wetland areas.

Real dependants
In the developing world, the rural economy and human well-being are even more closely dependent upon the wetland resource. Only rarely are national or household economies strong enough to replace goods or services once provided free by wetlands. The consequences of wetland loss are, therefore, fundamentally more severe in developing countries. There, loss of wetland resources leads not only to increased taxes, but to flood damage, contaminated water, human suffering and death.

▼ A camel at a waterhole in Sudan. As competition for the world's water resources increases, so does pressure upon the wetlands that depend upon that water. These pressures are likely to increase if global climate change causes higher temperatures and reduces rainfall in arid regions.

Similarly, in societies that rely on wetlands for fish protein, pasture, agricultural products or timber, any reduction in productivity is felt acutely. At best an increased proportion of the household budget has to be spent on subsistence and less on housing or education, while in many cases the reality is a lower-quality diet, or even a decline in total food intake. In the more extreme cases, as in many African floodplain systems, it can lead to rising mortality and emigration.

Rural development

In April 1989, a dispute over floodplain pastures in the Senegal Valley triggered ethnic violence which left over 1,000 people dead and tens of thousands homeless. While the ultimate reasons for this violence have a long and complex history, the immediate cause, a dispute over a valuable wetland resource, dramatically underlines the link between effective management of wetlands and human well-being. If well managed, wetlands can help meet people's needs; while their degradation and loss can worsen the already intense pressures upon rural communities in many parts of the world.

The traditional image of wetlands is one of an inaccessible waterlogged swamp that harbors disease-carrying mosquitoes. But times have changed. As more and more wetlands have been lost and others severely degraded we have grown to appreciate their uses. The challenge today is to develop approaches to wetland management which can ensure that these benefits are made available on a long-term basis and contribute most effectively to improving people's livelihoods.

Atlas of the World's Wetlands

Unlike many other endangered habitats, wetlands are not restricted to specific regions of the world. Their occurrence does not rely on direct rainfall, temperature or altitude, and for this reason examples of wetland systems are found in some of the hottest places on Earth as well as some of the coolest, from the breathtaking heights of the Andean Altiplano to Australia's Lake Eyre, which lies below sea-level. Wetlands are found everywhere, and a great many may soon be lost forever.

KEY TO ATLAS PAGES

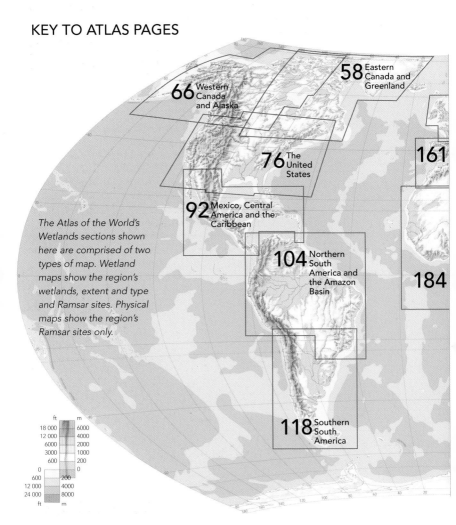

58 Eastern Canada and Greenland

66 Western Canada and Alaska

76 The United States

161

92 Mexico, Central America and the Caribbean

104 Northern South America and the Amazon Basin

184

118 Southern South America

The Atlas of the World's Wetlands sections shown here are comprised of two types of map. Wetland maps show the region's wetlands, extent and type and Ramsar sites. Physical maps show the region's Ramsar sites only.

ft	m
18 000	6000
12 000	4000
6000	2000
3000	1000
600	200
0	0
600	200
12 000	4000
24 000	8000
ft	m

Elevation and depth tints for Ramsar site maps

Height of land above sea level

in metres 6000 4000 3000 2000 1500 1000 400 200 0

Land below sea level
Depth of sea
in feet

6000 12 000 15 000 18 000 24 000

in feet 18 000 12 000 9000 6000 4500 3000 1200 600

0 200 2000 4000 5000 6000 8000 in metres

Some of the maps have different contours to highlight and clarify the principal relief features

Symbol key for Ramsar site maps

International boundaries

International boundaries
(undefined or disputed)

International boundaries show the de facto situation where there are rival
claims to territory

Internal boundaries

National parks

Perennial streams

Intermittent streams

Perennial lakes

Intermittent lakes

Swamps and marshes

Permanent ice
and glaciers

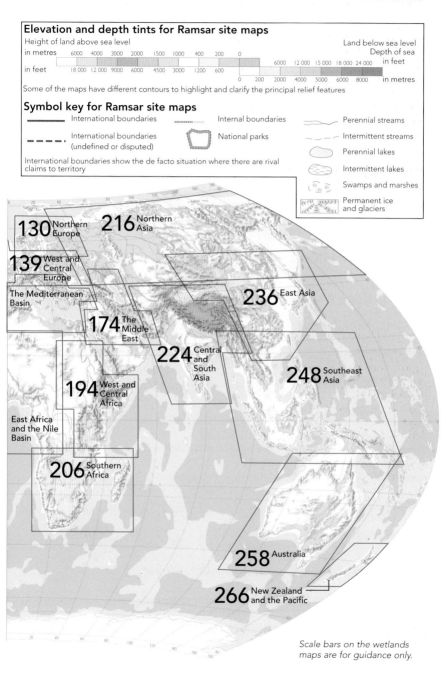

130 Northern Europe

216 Northern Asia

139 West and Central Europe

The Mediterranean Basin

236 East Asia

174 The Middle East

224 Central and South Asia

194 West and Central Africa

248 Southeast Asia

East Africa and the Nile Basin

206 Southern Africa

258 Australia

266 New Zealand and the Pacific

Scale bars on the wetlands maps are for guidance only.

57

Eastern Canada and Greenland

Eastern Canada is home to a diversity of wetland types that cover more than 220,000 square miles (570,000 square kilometers) – an area the size of France. These wetlands range from Arctic tundra ponds to boreal bogs and temperate fens; and from freshwater swamps to coastal salt marshes.

In some parts of eastern Canada, wetlands have come under severe pressure as a result of human settlement. In southern Ontario, since major settlement began in the mid-1800s, over 68 percent of the wetland areas has been converted to other land uses, either for agriculture, port or urban expansion. The wetlands around the Great Lakes have been among the most severely affected.

However, 70 percent of wetlands in eastern Canada are forested peatlands, lying in the boreal, more northerly, forest region. Here winters tend to be long and harsh, and summers short, and the peatlands have until recently experienced little development pressure. Concerns now exist concerning the pace and overall impact of forest development, oil and gas seismic exploration activities and the construction of hydroelectric reservoirs. This mosaic of forest, peatlands and other wetlands is a habitat of global importance, supporting millions of waterbirds, fish, beaver, moose, bears and other wildlife that depend on the boreal forest and its wetlands.

Conservation policies

Canada was one of the first countries to establish a national wetland policy. The Federal Policy on Wetland Conservation was approved by the Government of Canada in 1991. It has provided an important impetus for wetland conservation across the country. The policy has also served as an example that has been emulated by many other nations and has encouraged the development of complementary wetland policies in the majority of Canada's thirteen provinces and territories. This nation-wide effort and a broad cooperative approach to wetland management, has made Canada one of the world leaders in wetland conservation.

Protected sites

Over half of the wetland sites in Canada designated by the Ramsar Convention are located in the eastern provinces of the nation. They support a diversity of important wildife. Polar Bear Provincial Park on the northwest coast of James

Bay is important for the range of plants and animals that occur. These include not only the polar bear (*Ursus maritimus*), caribou (*Rangifer tarandus*), and many other species of mammals, but also hundreds of thousands of swans, geese and ducks. Overall, the Hudson and James Bay Lowlands support more than 4.5 million geese annually. The Hannah Bay and Moose River Migratory Bird Sanctuaries form the Southern James Bay Ramsar site. This is one of the most important staging areas in North America for migratory waterbirds that breed in the Arctic.

Further north on Baffin Island, the Dewey Soper Migratory Bird Sanctuary covers 3,150 square miles (8,160 square kilometers) of the Great Plain of the Koukdjuak. The Plain is an extensive Arctic oasis for wildlife, with sedge lowland and peaty deposits found throughout the area. It has a unique karst landscape of circular shallow ponds, raised beaches and shallow streams overlying limestone bedrock. A low relief shoreline results in ice-scoured tidal flats stretching 9 miles (15 kilometers) into the sea.

Designated a Ramsar site in 1982, the Plain supports the world's largest goose colony. In July and August, as many as 230,000 pairs of lesser snow geese (*Anser coerulescens*) nest here, some 30% of the Canadian

▼ *Canoeists paddle around a marsh at Point Pelee National Park in Ontario, Canada. The Park is one of the success stories of the Canadian conservation movement. In other parts of the country, however, stretches of similar coastal wetland are under threat from housing and recreational developments.*

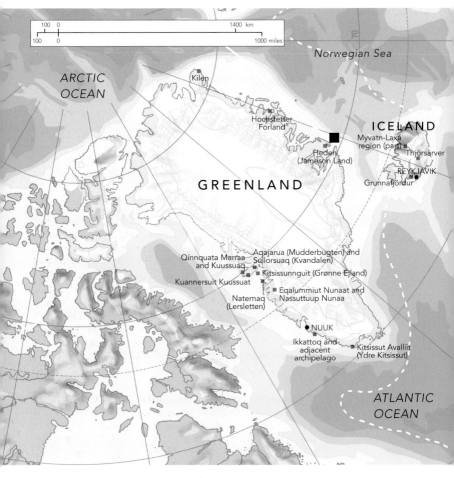

■ Ramsar Sites

population, and 10 percent of the world population. Canada geese (*Branta canadensis*), long-tailed duck (*Clangula hyemalis*), king eider (*Somateria spectabilis*) – the bill of which is considered an aphrodisiac in Greenland – and many shorebirds also nest in or use the area. The site is also an important habitat for one of Canada's main barren-ground caribou herds that migrates to and from the Plain each season.

The wetlands in the St. Lawrence Lowlands found on Canada's eastern coast include the popular Cap Tourmente National Wildlife Area near Quebec City. The Long Point National Wildlife Area and Point Pelee

■ Jameson Land

Jameson Land, a peninsula on the east coast of Greenland, includes a large area of lowland tundra area. Some 969 sq mi (2,524 sq km) of the tundra region has been designated a Ramsar site. Several rivers cut through the area and there are salt marshes on the coast. The many species of breeding birds found in the region include red-throated diver (*Gavia stellata*), barnacle goose, long-tailed skua (*Stercorarius longicauda*) and whimbrel (*Numenius phaeopus*). The largest number of birds, however, are nonbreeding geese, and 5,000 pink-footed and 2,500 barnacle geese come here to molt. Bulky, long-haired musk oxen (*Ovibos moschatus*) are abundant in this region and the molting geese are prey to the numerous arctic fox (*Alopex lagopus*).

National Park in southwestern Ontario are two other Ramsar sites in Canada visited frequently by the public. Along the St. Lawrence River, the Ramsar site located on the delta of Lac Saint Pierre protects some of the river's last remaining marshes. During the last 50 years 70% of these marshes have been drained, but Lac Saint Pierre remains as the largest freshwater floodplain in Quebec and supports 20,000 acres (8,000 ha) of marshes. Together with the surrounding farmland that is flooded each spring, these wetlands are one of the most important sites on the St. Lawrence for migratory waterbirds, and the lake is an important breeding site for herons. Some 1,300 pairs breed there annually.

Coastal and estuarine salt marshes are the wetlands under greatest pressure from development along Canada's Atlantic coast. Exemplified by several Ramsar sites, these marshes are critical areas for migratory shorebirds and waterfowl. The majority of all commercially harvested fish and crustaceans in Atlantic Canada depend upon these wetlands for part of their life cycle. Since settlement, about 65 percent of salt marshes in New Brunswick, Nova Scotia and Prince Edward Island have been altered or destroyed by the development of ports and harbors, or diking to permit agricultural expansion.

Greenland

The Greenland Icecap covers 80 percent of the island's land mass, with only parts of the coastal strip in the northeast and west remaining free in summer. This narrow strip is mainly mountainous, so the area of wetlands is small, and the deltas and estuaries are almost lifeless. The most important freshwater wetlands are usually a mixture of small lakes, marshes and moist tundra. In the

Ramsar Sites

Ramsar Sites/
Parks and Reserves

Water Bodies

25–50% Wetland

50–100% Wetland

Prairie Pothole Region

◀ *Greenland's summer landscape. When freed from ice during the long summer days, Greenland's coastal strip provides an important mixture of small lakes, marshes and moist tundra that are used by large numbers of breeding geese and breeding shorebirds. After raising their chicks on the invertebrates and plants that flourish here during the short growing season, the end of summer sees a mass exit as the birds head south to winter in temperate and tropical wetlands.*

Low Arctic zone, willow (*Salix* sp.) shrub is an important component. These inland wetlands provide the breeding habitat for six species of geese, one of which, the Greenland white-fronted goose (*Anser albifrons*), is an endemic subspecies. The High Arctic wetlands are also molting areas for pink-footed geese (*Anser brachyrhynchus*) and barnacle geese (*Branta leucopsis*), and nesting sites for large numbers of shorebirds, such as dunlin (*Calidris alpina*), sanderling (*C. alba*), knot (*C. canutus*) and ringed plover (*Charadrius hiaticula*).

The Great Lakes Wetlands Action Plan

Wetlands along the shores of the Great Lakes, particularly Lake Erie and Lake Ontario, are among the most threatened wetlands of both Canada and the United States. Studies of the shorelines of Lake Ontario since the early 1800s show that 90 to 95 percent of wetlands in this region have been converted to other land use needs.

Many shoreline swamps and marshes, and river delta marshes, have been developed into ports and harbors, industrial areas, diked for agriculture, or buried under other forms of urban development. Remaining wetland sites have been severely affected by toxic contaminants,

▲ Sanderling
(*Calidris alba*)

Queen Maud Gulf

Dewey Soper Migratory
Bird Sanctuary

N O R T H   W E S T
T E R R I T O R I E S

Southampton
Island

Baffin
Island

Hudson
Strait

McConnell River

H u d s o n
B a y

C A N A D A

MANITOBA

Polar Bear Provincial Park

James
Bay

Lake
Winnipeg

O N T A R I O

Quill Lakes

Last Mountain Lake

Southern James Bay
(Moose River and Hannah Bay)

Delta Marsh

Oak Hammock Marsh

Winnipeg

Lake Superior

0 km   500

0 miles   250

N

U N I T E D   S T A T E S

▲ Canada goose
(*Branta canadensis*)

▼ *Arctic fox (Alopex
lagopus) at Hudson Bay.
The Hudson-James Bay
coastal wetland is the
world's longest coastline
dominated by marshes.
In winter, it lies under
snow and ice, but after
the spring thaw, it turns
into a rich wetland.*

artificial barriers against the variable seasonal water levels, and by storm effects. Pressure to develop shoreline wetland sites for building vacation homes also remains high. Conservation of sites, such as the Point Pelee National Park and Long Point National Wildlife Area on Lake Erie, has been achieved through the cooperation of federal, provincial, state and local action, but many other ecologically important sites remain at risk.

Shepody Bay and Mary's Point

The extensive mud flats and tidal marshes at the head of the Bay of Fundy in New Brunswick, eastern Canada, form one of the most important resting and feeding areas for shorebirds in North America. This area has one of the largest tidal ranges in the world (up to 46 feet/14 meters) and the mudflats support the world's highest know density of the amphipod (*Corophium volutator*). Although common throughout European coastal areas, Corophium is found only in this region of North America and is the principal food source for the millions of migratory shorebirds that use the area. Four Ramsar sites are located here; Shepody Bay and Mary's Point together support 1 percent of the world population of semipalmated sandpiper (*Calidris pusilla*) and some roosting sites can support as many as 400,000 birds. Over a million

Ramsar Sites

Ramsar Sites/
Parks and Reserves

Water Bodies

25–50% Wetland

50–100% Wetland

sandpipers pass through Mary's Point between July and September each year. Other species using these sites include least sandpipers (*Calidris minutilla*), short-billed dowitchers (*Limnodromus griseus*), black-billed plovers (*Pluvialis squatarola*) and red knots (*Calidris canutus*). These habitats are irreplaceable – without them, sandpiper populations would not be able to store the fat and protein essential for their lengthy migration to overwintering havens in Suriname and other parts of South America.

Western Canada and Alaska

The landscapes of Alaska and the neighboring provinces and territories of western Canada contain most of North America's remaining wilderness areas. A rich diversity of subarctic, boreal, mountain, prairie and oceanic wetlands make up much of this wilderness. Taken together, western Canada and Alaska contain approximately 550,000 square miles (1.4 million square kilometers) of wetlands, an estimated 25 percent of the world's total.

Large estuarine deltas and associated salt marshes and tidal flats dominate the coastal wetlands. Some of the most extensive salt marshes lie along the 500-mile (800-kilometer) shoreline of the Alaskan Yukon-Kuskokwim Delta, itself one of the largest deltas in the world. To the seaward side of the marshes lie sand and mud flats, which are in places over 6 miles (10 kilometers) wide; in total they cover about 200 square miles (530 square kilometers).

The Copper River Delta, which lies just east of Prince William Sound, Alaska supports a variety of wetlands, including sandy barrier islands, estuarine tidal flats, salt marshes and freshwater marshes. The delta is an important staging area in the migration of numerous species of waterfowl and shorebirds. It is also a major nesting area for the dusky Canada goose (*Branta canadensis*) and for trumpeter swans (*Cygnus buccinator*).

1 Yukon-Kuskokwim Delta

The coastal plain of the Yukon-Kuskokwim Delta is one of the most productive wildlife areas in Alaska. It supports very high densities of water-fowl and shorebirds. In some areas, during late summer and fall, as many as 20 species of shorebirds use the intertidal zone, often found in densities exceeding 37 birds per acre (90 per hectare). Coastal estuaries and rivers are particularly important to the large flocks of molting birds and broods of brant, emperor, cackling and white-fronted geese (*Anser albifrons*).

2 Copper River Delta

In 1964, the entire Copper River Delta was lifted 6.5–13 ft (2–4 m) as a result of a massive earthquake which reached 8.5 on the Richter scale. Tidally influenced, sedge-dominated marsh has extended seaward by as much as 1 mile (1.5 km) across tidal flats, while preearthquake salt marshes have converted to freshwater systems dominated by shrubs and emergent plants.

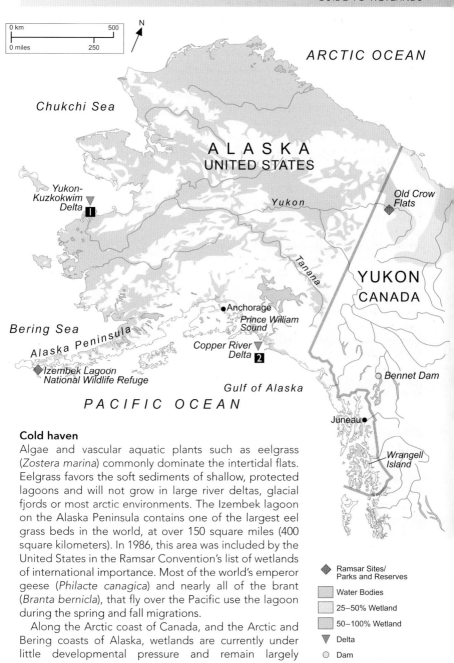

0 km 500
0 miles 250

N

ARCTIC OCEAN

Chukchi Sea

A L A S K A
UNITED STATES

Yukon-
Kuzkokwim
Delta ▼ **1**

Yukon

Old Crow
Flats ◆

Tanana

YUKON
CANADA

Bering Sea

Alaska Peninsula

●Anchorage
Prince William
Sound

Copper River ▼
Delta **2**

Bennet Dam ○

Izembek Lagoon ◆
National Wildlife Refuge

Gulf of Alaska

PACIFIC OCEAN

Juneau●

Wrangell
Island

Cold haven

Algae and vascular aquatic plants such as eelgrass
(*Zostera marina*) commonly dominate the intertidal flats.
Eelgrass favors the soft sediments of shallow, protected
lagoons and will not grow in large river deltas, glacial
fjords or most arctic environments. The Izembek lagoon
on the Alaska Peninsula contains one of the largest eel
grass beds in the world, at over 150 square miles (400
square kilometers). In 1986, this area was included by the
United States in the Ramsar Convention's list of wetlands
of international importance. Most of the world's emperor
geese (*Philacte canagica*) and nearly all of the brant
(*Branta bernicla*), that fly over the Pacific use the lagoon
during the spring and fall migrations.

Along the Arctic coast of Canada, and the Arctic and
Bering coasts of Alaska, wetlands are currently under
little developmental pressure and remain largely

◆ Ramsar Sites/
Parks and Reserves
☐ Water Bodies
☐ 25–50% Wetland
☐ 50–100% Wetland
▼ Delta
○ Dam

untouched. However, changes in economic and land use policy may open up the Arctic National Wildlife Refuge to oil exploration, with far-reaching consequences for the landscape. Along the Pacific coast, however, human populations are concentrated next to wetlands along the narrow coastal strip. In Juneau, Alaska, at least 13 percent of the wetlands have been filled for commercial and residential development since 1948.

Further south in Canada's Fraser River Delta, estuarine wetlands are also under intense development pressure. In winter, this delta supports the highest densities of waterfowl, shorebirds and birds of prey anywhere in Canada. In the summer, up to a million shorebirds of 24 species feed and rest in the delta. However, these wetlands compete with the needs of the urban center of Vancouver. From 1976 to 1982, over 28 percent of the wetlands in the Fraser lowland were developed for a variety of other uses.

Freshwater wetlands

Inland, freshwater boreal wetlands cover an estimated 230,000 square miles (600,000 square kilometers) of Alaska, and over 20 percent of central Canada. They include a wide range of wetland types, including peat bogs, fens, delta marshes, floodplain swamps and both wet tundra (which has areas of standing water) and moist tundra

▶ *An emperor goose (Anser canagicus) in its downy nest. A relative of the snow goose, the emperor goose breeds in western Alaska and northeastern Siberia, feeding among seaweed beds and estuarine mud flats.*

(with no areas of standing water). Peat is the dominant soil type in these wetlands, which contain a major portion of the world's carbon in the form of the extensive peat accumulations.

Moist and wet tundra underlain by permafrost (permanently frozen soil) cover vast areas in northern Canada and Alaska. These treeless landscapes are dominated by wetland grasses, sedges and dwarf shrubs. The common mare's tail (*Hippuris vulgaris*) and pendent grass (*Arctophila fulva*) are common in the wet tundra areas. Tussock cotton grass (*Eriophorum vaginatum*) and low shrubs such as dwarf birch (*Betula nana*) and bog blueberry (*Vaccinium uliginosum*) dominate the moist tundra. South of the tundra region and away from the coast, black spruce (*Picea mariana*) muskeg covers hundreds of thousands of square kilometers. Tamarack (*Larix laricina*) is associated with the black spruce in wet lowland sites. Common plants in the marshes of the region include water sedge (*Carex aquatilis*), marsh cinquefoil (*Potentilla palustris*) and bluejoint grass (*Calamagrostis canadensis*).

In the Pacific coast forest region, bogs and muskegs predominate where conditions are too wet for tree growth. The vegetation in these wetlands consists of Sphagnum mosses, sedges, rushes, shrubs and stunted lodgepole pine (*Pinus contorta latifolia*). Western hemlock (*Tsuga heterophylla*) or Alaska cedar (*Chamaccyparis nootkatensis*) dominate the forested wetland sites.

The majority of the freshwater wetlands of the region are subjected to little pressure from development. In Alaska, approximately 310 square miles (800 square kilometers) of freshwater wetlands have been lost since colonial times – about 0.1 percent of the original area. Until now, only a small area of Canadian boreal wetlands have been drained to improve forest productivity or developed for other purposes.

The North American Waterfowl Management Plan

The severity of wetland loss and the key role wetlands play in the western agricultural economy has become

◄ *Cotton grass (Eriophorum vaginatum) on Selawik Refuge Wetlands, Alaska. Cotton grass is common on the moist Alaskan tundra and provides an excellent spring time food source for animals such as the migratory porcupine caribou and their calves.*

widely recognized. As a result, Canada, the United States and Mexico have joined forces to create an innovative conservation program, the North American Waterfowl Management Plan, the aim of which is to return waterfowl populations to their 1970s levels by conserving wetland and upland habitats. Canada and the United States signed the plan in 1986 and Mexico joined in 1994. The plan, administered by the three national governments, is a partnership of federal, provincial/state and municipal governments, nongovernmental organizations, the private sector and many individuals working together to conserve and restore wetlands for birds and people. In this spirit, Joint Venture programs to conserve wetlands have been created under the plan in Alaska and all of Canada's provinces and territories. These are designed to both re-establish waterfowl and other wildlife populations, as well as promote soil and water conservation.

Achievements

As a result of its multipartner approach to conservation, and to over one billion Canadian dollars of funding from both Canada and the USA, the plan has been a major success. Over 5 million acres (2 million hectares) of wetlands and other migratory bird habitats have been conserved across Canada.

In southern Saskatchewan, the Missouri Coteau region is one part of the North American Great Plains that has retained a substantial area of natural habitat despite the pressures from agricultural development. Of a total area of over 5 million acres (2 million hectares), some 30 percent remains as natural habitat. The Coteau supports many pothole wetlands and is important for waterfowl and other wildlife, as well as Saskatchewan's cow-calf industry. Under the North American Waterfowl Management Plan, a range of partners are working with landowners to conserve and improve the condition of this remaining wetland habitat. For example approximately 600,000 acres (243,000 hectares) of the Coteau were given protection in 2004 through the work of Ducks Unlimited Canada, the Saskatchewan Watershed Authority, and the Nature Conservancy of Canada, working with local partners.

Beaufort Sea

UNITED STATES

Old Crow Flats

YUKON

Bennet Dam

PACIFIC OCEAN   Juneau

---

13 Ramsar Sites

13 Ramsar Sites are located in western and northern Canada, including the Queen Maud Gulf Migratory Bird Sanctuary that covers about 24,000 sq mi (62,800 sq km), an area almost the size of Ireland.

Ramsar Sites

Parks and Reserves

Ramsar Sites/ Parks and Reserves

Water Bodies

25–50% Wetland

50–100% Wetland

Delta

Dam

Boreal Wetlands

Canada's boreal forest stretches across the country from the border of Alaska down through the Yukon Territory, Northwest Territories, the north of British Columbia and the prairie provinces, and then across northern to Quebec, and Newfoundland and Labrador. Not only does this vast area hold about 25 percent of the world's remaining natural forests, but it is one of the world's most important reservoirs of freshwater held in the region's wetlands, lakes and rivers. These support tens of millions of breeding, staging and molting waterfowl and millions of shorebirds. The mosaic of wetland habitats that occur through the forest are also critically important for the bears, beaver, caribou, moose and wolves that use them, as well as for an estimated 3 billion land birds. Conservation of these habitats is emerging as a major priority.

71

Peace-Athabasca Delta

Stretching over 1,200 square miles (3,200 square kilometers), the Peace-Athabasca Delta in Alberta provides vistas of vast sedge meadows and shallow lakes. Their lush, emergent vege- tation provides an abundant food supply for wildlife, including the free-ranging bison (*Bison bison*) and a rich array of other mammal species such as gray wolves (*Canis lupus*) and lynx (*Felis lynx*). In spring and fall over a million birds, including many species of ducks and geese and the endangered whooping crane (*Grus americana*), use the delta.

Altered environment

Despite inclusion within Canada's biggest national park – Wood Buffalo – and a listing under the Ramsar and World Heritage Conventions, the Peace-Athabasca Delta is a striking example of the effects of external factors for even a large-scale wetland conservation initiative. In 1969, completion of the W.A.C. Bennet Dam for hydroelectrical storage vastly reduced water flows to the delta. As much as 25 percent of the delta has been affected and this could rise if there is any further dam construction or river diversion on the Peace River. These changes have been of special concern for the Fort Chipewyan aboriginal people, whose lives center on the delta. In the words of one resident, "Now the

▲ Bison (*Bison bison*)

▼ Lynx (*Felis lynx*)

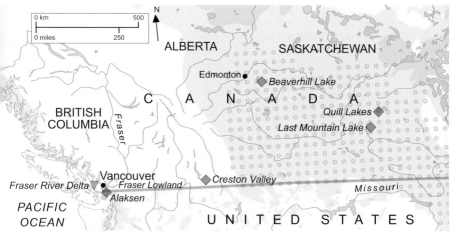

lakes where we used to trap are all poplar and willow. Over 100,000 to 300,000 muskrats came out of the delta every season. I used to get 50, 60, up to 100 muskrats a day. Only one year did I have to go on social assistance." Remedial actions have been taken however and weirs on the Peace River tributaries have nearly restored natural summer peak water flows in the delta, but the seasonal and annual fluctuations that are of such importance to maintaining the productivity and diversity of the delta are still less than under the natural regime.

Prairie potholes

The Prairies which stretch across the southern third of Alberta, Saskatchewan and Manitoba are a semiarid, generally treeless environment dotted with over 4 million small wetlands, known as potholes. This mosaic of wetlands continues to the south in the American states of Iowa, Minnesota, North Dakota, South Dakota and Montana. Referred to as the "duck factory" of North America, this area is critically important to breeding waterfowl. Of a total of 300,000 square miles (777,000 square kilometers), about 65 percent of the Prairie Pothole region lies in Canada; 106,000 square miles (274,000 square kilometers) are in the United States.

As the glaciers receded from this part of the continent 10,000 years ago, they left behind them small depressions in the landscape created by the scouring action of the ice. These "potholes" now support a variety of small wetlands, ranging from wet meadows and shallow water ponds, to saline lakes, marshes and fens. The vast majority of these wetlands collect water in the form of rain and melting snow, very few have surface inlets or outlets. The density of potholes may be as high as 155 sites every square mile (60 per square kilometer).

As rainfall is seasonal, many wetlands only have water for the spring and early summer periods. At this time they are flush with color as flowers bloom and birds breed. These wetlands play a vital role in the maintenance of nearly all forms of prairie wildlife, while also performing a wide range of other functions. Depending upon their location, basin shape and size, pothole wetlands can store floodwaters, absorb nutrients, recharge groundwater, and provide water and forage for domestic animals.

Agricultural pressure

The prairie landscape is, however, under intensive agricultural use for grain production. Cropland makes up about

Ramsar Sites/
Parks and Reserves

Water Bodies

25–50% Wetland

50–100% Wetland

Delta

Prairie Pothole Region

MANITOBA

Lake
Winnipeg

Oak-Hammock
Marsh
Winnipeg
Delta Marsh

68 percent of the prairie pothole region of the United States and Canada. Several technological events in modern history have spurred the agricultural development of the prairie lands. As early as 1930, the advent of the mechanized tractor eliminated the need to produce feed for mules and horses. Much of the rangeland that had been devoted to pasture was converted to cash crops with the result that between 1930 and the mid-1960s the total area of pasture land dropped from 100,000 square miles (263,000 square kilometers) to 11,000 square miles (28,000 square kilometers). Furthermore, since the 1960s, farm size, equipment size and the use of irrigation have all increased. These and other intensive farming practices have encouraged further drainage of wetland habitats and the use of uplands immediately adjacent to wetlands.

Vanishing potholes

The extent of prairie pothole wetlands has declined dramatically since settlement of the region in the 1850s. At that time, wetlands covered 16 to 18 percent of the prairie regions of the States of Minnesota, South Dakota, North Dakota and Iowa. Today, wetlands cover less than 8 percent of this region. In Iowa, the area of wetlands was reduced from more than 6,000 square miles (16,000

▼ A wintry, aerial view showing the prairie pothole terrain of North Dakota, USA. The "potholes", which vary dramatically in size, depth, shape and density, are the physical remains of the ice sheets that once covered this region of the United States and Canada. As the ice retreated, it scoured the land, creating thousands upon thousands of shallow depressions.

▲ Aerial photo showing wetlands and cropland in the prairie pothole region in northeastern South Dakota, USA. Modern farming equipment and methods have been responsible for the destruction on a massive scale of these wetland environments.

square kilometers) prior to 1875 to 1,450 square miles (3,800 square kilometers) in 1906, and 580 square miles (1,500 square kilometers) by 1922. Today in Iowa, those areas of wetland that still remain are only relicts of the former sites and most are protected under some form of state or federal stewardship.

In Canada, the patterns of drainage has been repeated since settlement of the Prairies in the early decades of this century. Many more small wetlands have been degraded by land use practices, such as water pollution from fertilizer and pesticide use, soil erosion and road construction. Together these practices have resulted in losses of as much as 71 percent of the original wetland area over the last 80 years.

The United States
– the lower "forty-eight"

When the British colonists established the first permanent settlement on the shores of Chesapeake Bay in 1607, the area that now makes up the lower 48 states comprised over 342,000 square miles (895,000 square kilometers) of wetlands. By 1997, the United States Fish and Wildlife Service (USFWS) estimated that only 165,000 square miles (427,000 square kilometers) remained. The processes of agricultural, industrial and urban development which dominated the four centuries since 1607 have been the major causes of American wetland loss.

Today, however, wetlands have become recognized as an invaluable public resource. Roughly two-thirds of the commercially important fish and shellfish species harvested along the Atlantic and Gulf coasts, and half of the Pacific coast species, are dependent on estuarine wetland habitats for food, spawning and/or nursery areas. And 60 to 70 percent of the 10 to 12 million waterfowl which nest in the lower 48 states do so in the Prairie Pothole region of the upper midwest. Millions of additional shorebirds, egrets, herons, terns, gulls, pelicans and other waterbirds depend upon a range of wetlands throughout the country.

The east coast of the United States presents a landscape of fresh- and saltwater marshes running into the estuaries dotted along the frequently highly indented coast. In the southeast, tidal wetlands are especially abundant in large, drowned river valley estuaries such as Delaware and Chesapeake bays. These estuaries are major resting and wintering grounds for the waterbirds that breed on the wetlands further north. They also provide a habitat for soft-shelled crabs, oysters and shrimp, and many fin fish, all of which yield important harvests and form an important part of the economy of the southeast.

Inland from the coast, the northeast is a land of freshwater bogs, fens and some extensive peatlands, with forested swamps as the single most important freshwater wetland. This pattern is mirrored in the unglaciated south. Here, extensive deciduous forested swamps, known as the "bottomland hardwoods", occupy the floodplains of river courses and large coastal basins which are flooded annually by slow-moving streams. The Great Dismal Swamp of Virginia and North Carolina, and the Okefenokee Swamp in Georgia are among the largest and best known. Freshwater, boglike wetlands called *pocosins* and Carolina bays are found on the coastal plain of the mid-Atlantic States.

Coast to coast

On the Florida Peninsula, freshwater originating from the north of Lake Okeechobee in central Florida flows south in wide, flat, shallow basins through the Great Cypress Swamp and the Everglades, the largest marsh system in the United States. The flow continues into the Gulf of Mexico, where fringing mangroves are North America's only forested marine wetlands. Around the Gulf Coast of Mississippi and Louisiana to the west, lie huge freshwater marshes and forested wetlands that have developed in the ancient deltas of the Mississippi River. These wetlands provide migratory and wintering habitats for many waterbirds, including several million ducks. In Texas, saline lagoons dominate much of the state's coastline.

Coastal marshes are less extensive along the Pacific coast and are located primarily in estuaries in the states of Oregon and Washington. Inland, the freshwater wetlands of the river basins and agricultural areas are important wintering areas for dabbling and diving ducks, geese and swans. However, these wetlands cover just a fraction of their former extent, and although California is the major waterfowl wintering area of the Pacific coast, the vast freshwater wetlands that once sat at the confluence of the Sacramento and San Joaquin rivers have been

▼ *Lower Klamath National Wildlife Refuge, Oregon, USA. Established in 1908, Lower Klamath NWR was the first national waterfowl refuge in the USA. The refuge has lost 80 percent of its original wetland extent.*

CANADA

Lake Superior

Sand Lake National
Wildlife Refuge

James

Missouri

Mississippi

Lake Michigan

Milwaukee
Horicon Marsh

Chicago

Platte

Missouri

St. Louis

Ohio

Cheyenne Bottoms
Quivira National Wildlife Refuge

Cache River-
Cypress Creek Wetlands

UNITED   STATES

Arkansas

Memphis

Mississippi

Cache-Lower White Rivers

Caddo Lake

Catahoula Lake

Rio Grande

Houston

New Orleans

Mississippi Delta

MEXICO

GULF OF
MEXICO

N

| 0 km | | 500 |
| 0 miles | 250 | |

Ramsar Sites

Ramsar Sites/
Parks and Reserves

Water Bodies

25–50% Wetland

50–100% Wetland

▽ Delta

Prairie Pothole Region

drained for agriculture. To the north in the Cascade mountain range, wetlands occur on narrow floodplains and on isolated alpine and subalpine meadows. Along the coast, tidal marshes fringe the rivers that flow into the Pacific Ocean. These wetlands once extended well inland, but are now diked for pastures. The estuary of the Columbia River and Puget Sound, however, still contain important tidal wetlands.

Lowland floodplains to glacial lakes

In the vast continent that lies between America's coasts, the wetland landscape changes from one of bogs, fens and forested swamps in the Great Lakes Basin to the Prairie Potholes of the upper Mississippi. Further south, wetlands occur primarily on the lowland floodplains and shifting river channels. Along the lower Mississippi, wetlands are inundated seasonally, while permanent wetlands occur in small floodplain ponds. These wetlands constitute the prime wintering area for waterfowl in the central part of the United States, and provide stop-over and resting areas in the single most important waterfowl migration route in North America.

To the west, wetlands are less abundant. In the Rocky Mountains, alpine wetlands are found on glacial lakes, and along slow-flowing streams in glaciated valleys. On the Columbia River plateau, wetlands are restricted to the river course, internal drainages, potholes and steep-sided lakes. Extensive marshes bordering the alkaline Harney and brackish Malheur lakes make up one of the most extensive inland marsh systems of the lower "forty-eight".

"No Net Loss"

Over the past 50 years wetland conservation has become an issue of national concern in the United States. Between the mid 1950s and the mid 1970s the USFWS estimated that 458,000 acres (185,400ha) of wetlands were lost each year. In the years up to the mid 1980s this had reduced to 290,000 acres (117,400ha) per year and in the period 1986–1997 was reduced further to 58,500 acres (23,700ha) per year. This drastic reduction in the rate of wetland loss is due to the combined efforts of federal and state agencies, and a wide range of national and local NGOs. Among the most successful actions taken at national level was the National Wetland Policy Forum that was established in 1987 by the Environment Protection Agency and the Conservation Foundation.

Based on an extensive process of consultation this forum recommended that the nation adopt the immediate goal of "no net loss" of wetlands and a long-term goal of actually increasing the nation's wetlands. While the continuing net loss of wetlands shows that the original goal of the forum has not been reached, the reduction in rate of loss is nevertheless a significant conservation achievement that reflects the efforts of many institutions and individuals.

Despite these achievements there is continued concern that pressures from various sectoral interests will lead to an upturn in the rate of wetlands loss in the future. Indeed the fragile nature of these recent successes is illustrated by a January 2001 decision of the US Supreme Court that is seen by many to jeopardize the wetlands protection provided by the US Clean Water Act. If these fears are realized, as much as 20 to 30 percent of the wetlands in the United States might be vulnerable to increased conversion pressure.

California's Central Valley

In the mid-1800s, the rivers of the California Central Valley flooded each winter to create vast seasonal wetlands in the valley floor and delta. In the 1850s, these wetlands amounted to more than 62,000 square miles (160,000 square kilometers). Vast flocks of migratory and resident waterbirds used these wetlands and rivers, which also provided spawning and rearing habitats for salmon and other species of fish. However, the 1850s saw the first widespread conversion of the wetlands when farmers diked the floodplains of the Sacramento Valley for cultivation.

Destructive activities

By 1939, 85 percent of the original wetland area had been lost, modified largely by levee and drainage activities, and local water diversion projects. Significant wetland losses have continued, with a further 8,000 square miles (20,000 square kilometers) of vegetated wetlands converted between 1939 and 1985. Agriculture was responsible for about 95 percent of the net loss of these wetlands. Today, only some 580 square miles (1,500 square kilometers) of all wetland types remaining in the Central Valley of California. The majority of these are managed areas that have been created or maintained by controlling water.

■ Ramsar Sites

◆ Ramsar Sites/ Parks and Reserves

Water Bodies

25–50% Wetland

50–100% Wetland

Salt Lakes

Prairie Pothole Region

CANADA

Alaksen

Puget
Sound

●Seattle

Columbia

Missouri

R
O
C
K
Y

Yellowstone

Snake

Harney Lake *Malheur Lake*

U N I T E D   S T A T E S

C
a
s
c
a
d
e

R
a
n
g
e

M
O
U
N
T
A
I
N
S

*Sacramento
Valley*

*Great
Salt Lake*

■*Tomales Bay*
■*Bolinas Lagoon*
●*San Francisco*

Denver●

■*Grasslands Ecological Area*

*San Joaquin
Valley*

◆*Ash Meadows
National Wildlife Refuge*

Colorado

●*Las Vegas*

PACIFIC

OCEAN

●*Los Angeles*

R
i
o

G
r
a
n
d
e

■*Tijuana River
National Esturine
Research Centre*

M E X I C O

0 km   500

0 miles   250

N

81

Chesapeake Bay

On the Atlantic coast of the United States, the Susque-hanna River ends at Chesapeake Bay. The largest estuary in north America, Chesapeake Bay drains 64,000 square miles and holds more than 18,000 billion gallons (82,000 billion liters) of water. It is one of the world's most productive large ecosystems and for centuries has produced large annual harvests of fish and shellfish. Historically for example the states of Maryland and Virginia harvested 17 million bushels of oysters, that produced as much as 110 million pounds (50 million kilograms) of meat.

By 1993 however the harvest of oysters in Maryland and Virginia had declined to 189,000 bushels. Similarly while harvest of blue crabs in the bay averaged 73 million pounds (33 million kilograms) between 1968 and 2003, the harvest in 2003 was only 48 million. These declines illustrate the steady decrease in environmental health of the bay over the course of the last century. The decline in shellfish is due to a combination of overfishing and degradation of the estuarine habitats on which they depend. In particular high nutrient inflows from agricultural and urban run-off have reduced oxygen levels in the bay during the summer.

Chesapeake Bay receives its freshwater from a 64,000-square mile (166,000-square kilometer) watershed that comprises six states. It extends from the head of the bay near Baltimore, in Maryland, through Pennsylvania to the

▼ *Early morning fog and fall colors reflected in the Patuxent River, one of the rivers draining into Chesapeake Bay, Mayland, USA – the largest estuary in the country. This immense system of fresh and saltwater marshes is highly productive, and some 90 percent of the nation's annual catch of striped bass (Roccus saxatilis) is taken from local waters, together with large numbers of crabs and shellfish. However, the long-term productivity of these wetlands is threatened by domestic pollution draining into the bay's river system.*

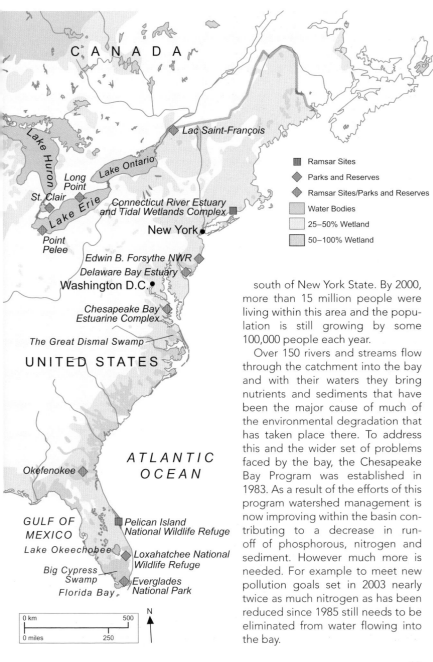

CANADA

Lac Saint-François

Lake Huron

Lake Ontario

Long Point

St. Clair

Lake Erie

Connecticut River Estuary and Tidal Wetlands Complex

New York

Point Pelee

Edwin B. Forsythe NWR

Delaware Bay Estuary

Washington D.C.

Chesapeake Bay Estuarine Complex

The Great Dismal Swamp

UNITED STATES

ATLANTIC OCEAN

Okefenokee

GULF OF MEXICO

Pelican Island National Wildlife Refuge

Lake Okeechobee

Loxahatchee National Wildlife Refuge

Big Cypress Swamp

Everglades National Park

Florida Bay

Legend:
- Ramsar Sites
- Parks and Reserves
- Ramsar Sites/Parks and Reserves
- Water Bodies
- 25–50% Wetland
- 50–100% Wetland

0 km — 500
0 miles — 250

N

south of New York State. By 2000, more than 15 million people were living within this area and the population is still growing by some 100,000 people each year.

Over 150 rivers and streams flow through the catchment into the bay and with their waters they bring nutrients and sediments that have been the major cause of much of the environmental degradation that has taken place there. To address this and the wider set of problems faced by the bay, the Chesapeake Bay Program was established in 1983. As a result of the efforts of this program watershed management is now improving within the basin contributing to a decrease in run-off of phosphorous, nitrogen and sediment. However much more is needed. For example to meet new pollution goals set in 2003 nearly twice as much nitrogen as has been reduced since 1985 still needs to be eliminated from water flowing into the bay.

Mississippi Delta and the bottomland hardwoods

The southern states of the United States support some of the country's most important wetland resources. The Mississippi Delta alone contains about 30 percent of all coastal wetlands in the lower 48 states. Equally important, but more widespread in their distribution, are the bottomland hardwood wetlands which characterize the floodplains flanking the major streams of the southeast of the country.

As in many other regions of the world, these wetlands have seen many changes over thousands of years, but in recent decades the changes have become more and more dramatic. For example, over the past 7,000 years, a total of 14,000 square miles (36,000 square kilometers) of land have been built up in the Mississippi Delta, while some 7,000 square miles (18,000 square kilometers) of this land have in turn been washed away as the river's path has changed as a result of natural river processes.

▼ *Flood defenses along the banks of the lower Mississippi have increased water velocity so that sediment that was once deposited in the delta is now carried out to deep water. As a result, the delta is shrinking rapidly, at a rate of about 40 sq mi (100 sq km) per year. If remedial action is not taken, the Gulf coastline will look very different in 40 years' time (shown by the green line).*

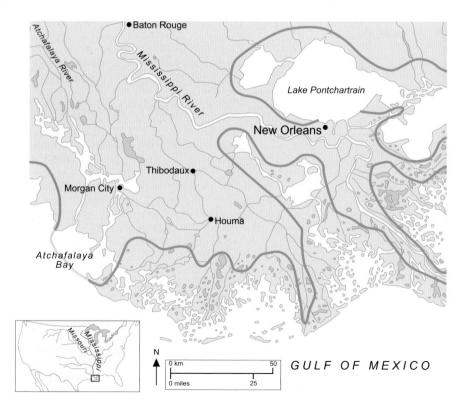

▶ *The muskrat (*Ondatra zibethicus*) is one of the many animals that inhabit the wetlands around the mouth of the Mississippi River, and these traditional habitats are now threatened.*

Gulfort

Mississippi Delta

Today however there is an annual net loss of some 35 square miles (90 square kilometers) of the delta being due primarily to the leveeing of the river. In the past, the sediment carried by the river was deposited in the delta, but the modified river now carries this far offshore into the deep water of the Gulf of Mexico.

In the lower Mississippi valley, more than 1,500 miles (2,500 kilometers) of dikes have been built and the river is confined for more than 600 miles (1,000 kilometers) of its course. In addition to shortening the river, increasing stream velocities and contributing to the erosion of the delta, this engineering work now prevents natural flooding of the floodplain forested wetlands. These once covered 39,000 square miles (100,000 square kilometers), but now only 7,000 square miles (18,000 square kilometers) remain; the rest was cleared to provide land for agriculture. A total of 1,900 square miles (900 square kilometers) of wetlands have been lost in Louisiana since the 1930s and another 700 are projected to disappear by 2050.

It was the desire for economic development that brought about these large-scale developments. However, the economic costs, along with the investments required to maintain the Mississippi Delta, are only now being widely realized. It may already be too late for many of the larger mammals, such as jaguars (*Panthera onca*) and black bears (*Ursus americanus*), whose populations in these last refuges have been severely diminished. However increasing attention is now being given to these issues and in 2004 the state of Louisiana proposed a US$14 billion plan to save the remaining wetlands. This highlights the importance of recognizing the value of wetlands at an early stage in national or regional development. Few other countries will be able to invest anywhere close to this sum to repair the damage caused by damaging land-use policies.

"Reclaiming" land for agriculture

While the Mississippi symbolizes the impact of a century of intensive development upon the nation's wetland resources, many other wetland systems are equally threatened. Less than 35 percent of the bottomland hardwood forest which existed when white settlers arrived now remain. Again, the major force behind this loss has been agricultural development. Between the 1950s and 1970s, Louisiana, Arkansas and Mississippi lost vast areas of bottomland hardwoods to crop production, primarily soyabean and cotton. Agricultural subsidies encouraged wetland loss, and much of the money for draining and diking of the bottomland hardwoods came from government subsidies. As awareness of the importance of wetlands grew, together with concern for their conservation, steps were taken to halt this destruction of a unique resource. Over the past two decades systems of multiple use have been promoted, which make maximum sustainable use of the timber, fish and shellfish, wildlife and recreation values of the wetlands without affecting the valuable hydrological functions, such as water purification and flood retention. As a result of these and other measures the rate of wetland loss has slowed significantly, although 1,200,00 acres (486,000ha) of freshwater forested wetlands were still lost in United States between 1986 and 1997.

▼ Waterfowl fly over Lacassine National Wildlife Refuge, Louisiana. Lacassine refuge supports one of the largest concentrations of waterfowl in the US. The state of Louisiana has pledged US$14billion in a bid to save the state's vanishing wetlands.

▲ *Hardwood wetlands, Georgia, USA. This floodplain cypress forest is a haven for many different species of wildlife. The introduction of flood control levees can have a devastating effect on this special habitat.*

The productive hardwoods

The bottomland hardwood wetlands, characterized by bald cypress trees and still, duckweed-covered waters, once stretched over 140,000 square miles (370,000 square kilometers), but today cover less than 50,000 square miles (130,000 square kilometers). These breathtaking wetland systems support plant and animal communities that are highly adapted to fluctuating water levels and are closely coupled to the estuarine systems at the river mouths.

Such floodplain forests are highly productive. The annual floods supply the system with sediments, dissolved organic matter and nutrients. Over 50 species of fish feed or spawn on the floodplains, and their distribution and abundance in the rivers is dependent on the forests' flooding cycle. When flood control levees are built along the stream banks to prevent flooding and encourage conversion of the forest to other uses, this critical breeding habitat is lost.

At the mouths of many of the rivers that drain the bottomland hardwoods, there are estuaries that support highly productive and valuable marine fisheries. The same flooding that brings life-supporting elements to the floodplain forests also flushes organic material and silt from the forest floor and delivers it to the estuaries. This flushed material, some of which comes from the leaf and twig litter of the forests themselves, supports the estuaries' complex food chains. In this way, the annual flood is an essential contributor to the economic viability of the valuable fish and shellfish industry of the river mouths.

The Everglades

The Florida Everglades comprises one of the largest freshwater marshes in the world. Once this wet wilderness covered 4,000 square miles (10,000 square kilometers), from Lake Okeechobee to the southwest tip of Florida. At that time the Everglades formed what the Indians who lived there called the "grassy waters", a river of grass creeping to the sea.

As the name suggests, the Everglades is dominated by sawgrass (Cladium sp.), which can form single stands covering hundreds of square yards and often reaching up to 13 feet (4 meters) in height. The Everglades also supports a great variety of aquatic plant communities, including lilies (Nuphar and Nymphaea spp.) and bladderworts (Utricularia sp.), which mix with the sawgrass in many areas. Where irregularities in the limestone bedrock have left raised areas, small islands or hummocks develop. These hummocks are populated by a diversity of hardwood trees and support varied populations of animals, including the Liguus tree

▶ Red mangroves (Rhizophora mangle), Everglades National Park. The mangroves' "stiltlike" prop roots are one of many adaptations to living in the soft substrate of the coastal fringe. The transition from the freshwater forest and meadows of the Everglades out to the marine habitats of Florida Bay is one factor that contributes to the diversity of wildlife that the Everglades supports.

◄ *The wildlife diversity of the Everglades.*

1 Everglade Kite,
   Rosthamus sociabilis
2 Pileated woodpecker,
   Phloeceastes pileatus
3 Prothonotary warbler,
   Protonotaria citrea
4 Rough green snake,
   Opheodrys aestivus
5 Limpikin,
   Aramus guarauna
6 Racoon, *Procyon lotor*
7 Flamingo,
   Phoenicopterus ruber
8 Mississippi alligator,
   Alligator mississippiensis
9 Roseate spoonbill,
   Ajaia ajaja
10 Eastern fox squirrel,
   Sciurus niger
11 Snail,
   Isognomon melina alata
12 Green treefrog,
   Hyla cinerea
13 Green anolis,
   Anolis carolinensis
14 Zebra butterfly,
   Heliconius charitonius
15 Purple gallinule,
   Porphyrula martinica
16 Snail, *Pomocea flagellata*
17 Swamp rabbit,
   Sylvilagus sp.
18 Oxeye tarpon,
   Megalops cyprinoides
19 Cottonmouth moccasin,
   Ancistrodon piscivorus

snails, which are endemic to southern Florida and the islands of Cuba and Hispaniola.

North of the marsh lies Big Cypress Swamp, while to the south lie the fringing mangroves of Florida Bay. These forested wetlands complement the sawgrass and its associated communities, and provide nesting habitats for the colonial waterbirds that feed in the marsh. Alligators, on the other hand, confine themselves to the freshwater marshes, with the more rare American crocodiles living in salty and brackish waters.

The Miami Canal

For much of the nation's early history the Everglades was untouched by major development efforts. However, at the turn of the century Florida's governor initiated the first drainage project, which was completed in 1909. Known as the Miami Canal, it connected Lake Okeechobee to the Miami River and the sea. Severe flooding in 1926, 1928, 1947 and 1948 resulted in the Central and South Florida Control project, which involved the construction of almost 800 miles (1,300 kilometers) of levees and 500 miles (800 kilometers) of canals, further helping to drain the land.

Today, less than 50 percent of the original Everglades remains, the rest has been drained and converted to agricultural or urban uses. The freely flowing "river of grass" has been replaced by an intensively managed,

◀ West of the Everglades lies Big Cypress Swamp, which comprises 4,000 sq km (1,500 sq miles) of primeval glory. Between the towering bald cypress (Taxodium distichum), a relative of the California redwood, are clusters of cypress "knees" which probably serve to keep the tree roots aerated. Although it is dominated by bald cypress, Big Cypress Swamp also supports other tree species, including pond dwarf cypress (T. ascendens), as well as slash pine, pond apple and willow on drier ground. Big Cypress is an important wildlife conservation area, and it is thought to contain some of the country's last remaining black bears.

multiple-purpose water control system containing over 1,900 miles (3,000 kilometers) of canals and levees, and 150 major water-control structures. Only a fifth of the original Everglades, the most downstream portion, is protected within the Everglades National Park. This remnant of the original system, however, faces a variety of threats from upstream. The quantity, quality and distribution of water flowing through the Everglades, all of which are critical to maintaining the ecological character of the Everglades, has been altered and biodiversity reduced. For example many of the plant species that occur in the Everglades are adapted to the low-nutrient content of the waters that naturally flow through the system. When the nutrient content of these waters is increased as a result of urban or agricultural run-off many the original plant species such as spike rushes (Eleocharis) and bladderworts (Utricularia) and the communities they support are replaced by grasses (*Panicum*) and arrow-head (*Sagittaria*).

Hope for the future

The challenge that the Everglades now presents to scientists, engineers, planners and managers is whether or not it can coexist with, and be part of, an intricate public water-management system – one that has to accommodate the unchecked water supply and flood control needs of agriculture, economic growth, and south Florida's expanding human population. Thus, while much has been achieved in the past few decades, starting with the Florida State Government's multimillion dollar "Save our Everglades" program in 1983 and the injection of millions of dollars of federal funds, the battle continues on.

In 1988, the Federal Government sued the State of Florida and the South Florida Water Management District over its failure to prevent degradation of wetland ecosystems in the Loxahatchee National Wildlife Refuge and the Everglades National Park. This led to Florida's Everglades Forever Act in 1994 and in turn to the establishment of a restoration plan. Building on these steps further calls for increased conservation investment led in 2000 to congressional approval of the US$8 billion "Comprehensive Everglades Restoration Plan". Scheduled to take over 30 years to complete, this is the largest environmental restoration project to be undertaken in the United States.

▼ *Great white egret* (Egretta alba) *fishing in the Everglades.*

Full implementation of the plan will require a sustained commitment by State and Federal governments to provide the funding, while also preventing other actions that might accelerate environmental degradation before the plan can be completed. For example in 2003 the State of Florida postponed the introduction of higher water quality standards from 2006 to 2016, so raising concern that the water quality of the Everglades will continue to decline. The National Wildlife Federation has also drawn attention to the fact that while the Everglades Restoration Plan is focused on the eastern and southern parts of Florida, drainage of wetlands is continuing in the western Everglades. Much more conservation investment is clearly required if these areas are also to be protected.

Mexico, Central America and the Caribbean

The natural value of Mexico's water and wetland resources has influenced both the country's history and the fortunes of its people. The ancestors of the Aztecs are thought to have emigrated from Aztlan ("the place of the white herons") along the Pacific coast of Mexico. They established a new city, Meico-Tenochtitlan, on a volcanic island surrounded by the waters of the ancient lakes of Chalco and Texcoco. Their agriculture was based on cultivating vegetables, fruit, flowers and herbs on floating islands made out of reed mats and mud.

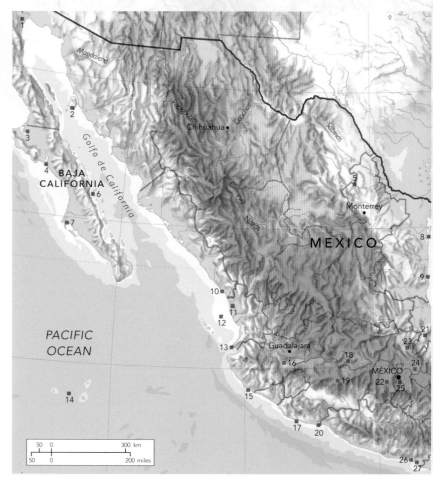

■ **Ramsar Sites**

1 Humedales del Delta del Río Colorado
2 Isla San Pedro Mártir
3 Laguna Ojo de Liebre
4 Laguna San Ignacio
5 Área de Protección de Flora y Fauna Cuatrociénegas
6 Parque Nacional Bahía de Loreto
7 Laguna Playa Colorada – Santa María La Reforma
8 Laguna Madre
9 Playa Tortuguera Rancho Nuevo
10 Playa Tortuguera El Verde Camacho
11 Marismas Nacionales
12 Parque Nacional Isla Isabel
13 Islas Marietas
14 Reserva de la Biosfera Archipiélago de Revillagigedo
15 Reserva de la Biosfera Chamela-Cuixmala
16 Laguna de Sayula
17 Playón Mexiquillo
18 Laguna de Yuriria
19 Humedales del Lago de Pátzcuaro
20 Laguna Costera El Caimán

21 Presa Jalpan
22 Ciénegas de Lerma
23 Laguna de Metztitlán
24 Laguna de Tecocomulco
25 Sistema Lacustre Ejidos de Xochimilco y San Gregorio Atlapulco
26 Playa Tortuguera Tierra Colorada
27 Playa Tortuguera Cahuitán
28 La Mancha y El Llano
29 Sistema de Lagunas Interdunarias de la Ciudad de Veracruz
30 Parque Nacional Sistema Arrecifal Veracruzano
31 Cuencas y corales de la zona costera de Huatulco
32 Sistema Lagunar Alvarado
33 Manglares y humedales de la Laguna de Sontecomapan
34 Reserva de la Biosfera Pantanos de Centla
35 Parque Nacional Cañón del Sumidero
36 Reserva de la Biosfera La Encrucijada
37 Parque Nacional Lagunas de Montebello
38 Áreas de Protección de Flora y

Fauna de Nahá y Metzabok
39 Área de Protección de Flora y Fauna de Términos
40 Playa Tortuguera Chenkán
41 Reserva de la Biosfera Los Petenes
42 Reserva de la Biosfera Ría Celestún
43 Reserva Estatal El Palmar
44 Dzilam (reserva estatal)
45 Humedal de Importancia Especialmente para la Conservación de Aves Acuáticas Reserve Ría Lagartos
46 Área de Protección de Flora y Fauna Yum Balam
47 Parque Nacional Isla Contoy
48 Parque Nacional Arrecife de Puerto Morelos
49 Playa Tortuguera X'cacel-X'cacelito
50 Parque Nacional Arrecife de Cozumel
51 Laguna de Chichankanab
52 Sian Ka'an
53 Bala'an K'aax
54 Reserva de la Biosfera Banco Chinchorro
55 Parque Nacional Arrecifes de Xcalak

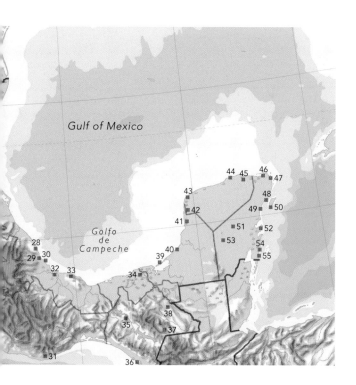

Gulf of Mexico

Golfo de Campeche

During the colonial period, the historical links with the coast were weakened, and the European immigrants concentrated upon inland agriculture, mining and lumber harvest. By heading inland, the immigrants avoided the hot, humid, disease- and insect-plagued coastal areas. Although this development placed intense pressure on inland wetlands – one of the "conquistadores'" main concerns was to drain the lakes surrounding Tenochtitlan – the coastal areas remained largely untouched. Since the 1980s, however, Mexico has seen increased coastal developments, mainly in aquaculture, offshore oil exploitation and tourism, all of which have contributed to wetland degradation and loss.

Wetland wealth

Today Mexico has some 2,500 square miles (6,500 square kilometers) of inland wetlands, mainly lakes, lagoons and rivers. The country's main wetlands, however, still lie on the coast, covering an area of some 4,800 square miles (12,500 square kilometers). Many of Mexico's wetland species are rare or endangered, notably freshwater and marine turtles, manatee, caiman, crocodiles, several

▼ *Lying on the northwestern tip of the Yucatán peninsula the Biosphere Reserve of Ría Celestún is one of Mexico's most important coastal wetlands and is listed as a Ramsar site. It contains a mix of wetland habitats including large areas of mangroves and flooded forest. This ecological mosaic supports over 300 bird species.*

species of fish and many resident and migratory birds. Waterbirds are especially well studied, and critical sites well documented. The Pacific coast is especially important for wintering brown pelican (*Pelecanus occidentalis*), brent goose (*Branta bernicla*), great blue heron (*Ardea herodias*), and many species of shorebird.

The major threats to Mexico's wetlands are agricultural and industrial development. Yet coastal lagoons and estuaries, which cover 6,000 square miles (16,000 square kilometers), are capable of producing food equivalent to that produced on 62,000 square miles (160,000 square kilometers) of Mexico's agricultural land. Lagoons such as Alvarado, Terminos and the wetlands in Tabasco in the Gulf of Mexico, and Huizache-Caimanero and the coastal areas of Nayarit in the Pacific, have the highest productivity of any habitat type in Mexico. If adequately managed they could in a year produce close to 160 pounds of oysters per acre (180 kilograms per hectare), over 10 times the amount of beef produced on drained wetlands. Yet wetland loss continues at an alarming rate.

The Caribbean

The Insular Caribbean is generally considered to consist of some 23 nation states, spread over 3,000 individual islands, cays and islets. The majority border the Caribbean Sea, although the Bahamas, Turks and Caicos Islands, and Bermuda to the northeast are included with the West Indies even though they lie outside the main archipelago in the Atlantic Ocean.

Approximately 91,000 square miles (235,000 square kilometers) of wetlands are found on these islands. The wide range in island size, topography and local climatic conditions have created immensely varied wetland types. The most extensive are found in the large islands of the Greater Antilles, the French Antilles and Trinidad, with the largest being the Ciénaga de Zapata in Cuba, which covers 1,300 square miles (3,400 square kilometers). Generally, these are high islands, with much of their central land area over 3,300 feet (1,000 meters). Many support rivers with floodplain and estuarine environments in which complex wetland systems can develop. In contrast, in low islands, such as Anegada, the Bahamas and the Caicos Islands, where the height of the land rarely exceeds 100 feet (30 meters), there are few rivers and wetlands are restricted to saline coastal types.

Freshwater marshes are rare in the Caribbean. This is because some have been converted for agriculture,

particularly for rice cultivation, but also because the climate favors swamp forests over herbaceous aquatic plants. Nonetheless, there are some large marshes in the greater Antilles and Trinidad and most of the high eastern Caribbean islands have small patches.

Very few Caribbean wetlands are legally protected or managed for resource use. The larger sites have received the most attention and several of them are reserves or are included in reserves, such as the Zapata National Park, which incorporates the Ciénaga de Zapata. In general, however, there is widespread degradation of mangrove areas by cutting for charcoal burning and coastal pollution, waste dumping and land reclamation, while many freshwater sites also suffer from encroachment for farming coupled with burning.

Central America

The Central American isthmus connects the continental masses of North and South America, and separates the Caribbean from the Pacific. In spite of its small size, only 190,000 square miles (500,000 square kilometers), it possesses an extraordinary biological, physical and cultural diversity. Wetlands are an important part of Central

▼ *Brown pelican* (Pelecanus occidentalis), *left and Great blue heron* (Ardea herodias), *right.*

Ramsar Sites

Ramsar Sites/
Parks and Reserves

Water Bodies

Peatlands

America. Freshwater swamps, peatlands, swamp palm forests, floodplains, coastal lagoons, and mangrove forests are common features and support a wide range of uses by local communities.

The central mountain chain is a strong influence on the climate of the region. It acts as a barrier to wind circulation, and makes the Caribbean much more humid than the Pacific. Close to 70 percent of the region's total rainfall drains into the Caribbean, where river basins are larger, water volumes higher and river flow fairly regular throughout the year. Rivers on the Pacific side are shorter, with less regular and lower water volumes, and a distinct dry season. On the Pacific coast, mangrove forests and temporary nontidal wetlands are most common; while on the Caribbean side the variety of wetlands is greater, including permanent and temporary freshwater marshes, palm swamps and freshwater swamp forests, forested peatlands, and important coastal lagoons, mangroves, coral reefs and seagrass beds.

British Virgin Islands

Western Salt Ponds of Anegada

Puerto Rico

Guadeloupe

Grand Cul-de-Sac Marin de la Guadeloupe

French Antilles

Savannes Bay/ Mankoté Mangrove

ST. LUCIA

ST. VINCENT

BARBADOS

GRENADA

TRINIDAD AND TOBAGO

Port of Spain

Nariva Swamp

Population pressure

While climate and hydrology explain the distribution of the different types of wetlands in the region, demography and history help explain the pattern of wetland use. During colonial times, between 1500–1700 population growth was slow; even by the early 1900s only about 3 million people inhabited Central America. By 2001, however, the population had grown to 37 million, with 39 percent less than 15 years old. It is estimated that by the year 2015 there will be some 49 million Central Americans.

Most of the population is concentrated in the central highlands and along the Pacific coast where there is now severe deforestation, erosion and depletion of natural resources. In recent decades, this pattern of resource loss has led to emigration into the extensive and humid Caribbean plains where the new settlers have brought with them their traditional production systems from drier areas. As a result, slash-and-burn agriculture and extensive cattle ranching have led to erosion of the fragile soils.

Ramsar Sites

Ramsar Sites/
Parks and Reserves

Water Bodies

Coastal Wetlands

Freshwater Marsh/
Floodplains

Mangroves

Swamp/Flooded Forest

Peatlands

Salt Pans

1 Dzilam
2 Yum Balam
3 Parque Nacional Isla Contoy
4 Parque Nacional Arrecife de
   Puerto Morelos
5 Playa Tortuguera X'cacel-
   X'cacelito
6 Sian Ka'an
7 Banco Chinchorro
8 Parque Nacional Arrecifes de
   Xcalak
9 Laguna de Chichankanab
10 Bala'an K'aax
11 El Palmar
12 Ría Celestún
13 Los Petenes
14 Playa Tortuguera Chenkán
15 Laguna de Términos
16 Pantanos de Centla
17 Nahá y Metzabok
18 Parc Natcional Lagunas de
   Montebello
19 Crooked Tree Wildlife
   Sanctuary

20 Parc Natcional Laguna del
   Tigre
21 Manchón-Guamuchal
22 Refugio de Vida Silvestre
   Bocas del Polochic
23 Punta de Manabique
24 Parque Nacional Jeanette
   Kawas
25 Refugio de Vida Silvestre
   Punta Izopo
26 Barras de Cuero y Salado
27 Laguna de Bacalar
28 Sistema de Humedales de la
   Zona Sur de Honduras
29 Laguna del Jocotal
30 Deltas del Estero Real y Llanos
   de Apacunca
31 Lago de Apanás-Asturias
32 Cayos Miskitos y Franja
   Costera Immediata
34 Sistema de Humedales de San
   Miguelito
33 Sistema Lagunar de Tisma
35 Sistema de Humedales de la

Bahía de Bluefields
36 Caño Negro
37 Manglar de Potrero Grande
38 Laguna Respringue
39 Los Guatusos
40 Tamarindo
41 Palo Verde
42 Refugio de Vida Silvestre Río
   San Juan
43 Cuenca Embalse Arenal
44 Humedal Caribe Noreste
45 Turberas de Talamanca
46 Gandoca-Manzanillo
47 Terraba Sierpe
48 San San-Pond Sak
49 Golfo de Montijo
50 Bahía de Panamá
51 Punta Patiño

Today, wetland degradation on the Caribbean side continues with the expansion of the agricultural frontier, particularly for banana, citrus fruit and oil palm plantations. Drainage of freshwater wetlands is now common, and siltation, eutrophication and high levels of pesticides are affecting the previously pristine coastal lagoons of the Caribbean coast. Lately, tourism development and construction of recreational housing have added to the pressures for wetlands drainage.

Along the Pacific, where almost all of the dry forests have been lost, mangroves have become an important source of firewood. This is used for both domestic and industrial purposes, including salt production. While such practices yielded substantial benefits for the population, it is questionable whether mangroves in such areas as the Pacific coast of Nicaragua can sustain these current high levels of exploitation. Shrimp mariculture is also a major threat to the mangroves of the region. In the 1980s concessions of mangrove land for mariculture in the Gulf of Fonseca in Honduras cost less than US$0.40 per acre per year (US$1 per hectare) for the first three years. As a result most shrimp ponds were constructed in cleared mangrove forests, even though it has been demonstrated elsewhere that operational costs are lower and production higher in other coastal sites.

These pressures have combined to decrease substantially the area of mangroves. For example, Guatemala lost half of its mangroves between 1960 and 1985. In Panama more than 150 square miles (400 square kilometers) of the mangroves have been lost to agriculture, ranching and shrimp mariculture.

▼ *Aerial spraying of a banana plantation in Bocas del Toro, Panama. Banana plantations have damaged wetlands along Central America's Caribbean coast. Plants cannot survive periods of waterlogging, and the draining canals that are created to keep the plants alive carry the large quantities of fungicides, fertilizers and pesticides used directly into many rivers.*

◄ *Manatee* (Trichechus manatus). *A peaceful herbivorous animal that feeds on plants growing on the seafloor or riverbed. Lives in fresh or brackish estuarine waters.*

Conservation policies

In the face of these pressures, the degree of protection offered to wetlands in Central America differs in each country. In response to these all seven countries have signed the Ramsar Convention, and a total of 33 sites have been designated as Wetlands of International Importance. This is a striking demonstration of the growing awareness of the importance of wetlands in the region and the need for increased conservation efforts.

Each country in the region has created its own Protected Areas System and important wetlands have been included, such as Crooked Tree Wildlife Sanctuary in Belize, Los Guatusos Wildlife Refuge in Nicaragua, Chocôn-Machacas Biotope in Guatemala, Palo Verde National Park in Costa Rica and Cuero y Salado Wildlife Refuge in Honduras. Although these wetlands were declared protected areas mainly because of the important wildlife they harbor – crocodiles and caimans, manatees and migratory and resident waterfowl – the close linkages between human populations and wetlands in Central America are now widely realized. If wetland management efforts in the region are to be successful, local populations must be involved and receive benefits from this.

El Salvador's Laguna El Jocotal, a Ramsar site, provides an excellent example of the way in which people benefit from wetland conservation. Lying in a volcanic crater, Jocotal is a freshwater lagoon which varies in size from 2 square miles (5 square kilometers) in the dry season to

6 square miles (15 square kilometers) in the rainy season. More than 130 species of aquatic birds use the lagoon, with the tree duck *Dendrocygna autumnalis* occurring in large numbers. By 1977, hunting, egg poaching, and loss of nest sites – a consequence of deforestation around the shores of the lagoon – had combined to reduce the population of this particular species to less than 500 individuals. In response, the Salvador National Park Service began a collaborative program with the local communities to control hunting and install nest boxes. In return, locals were allowed to collect a proportion of the eggs from some of the nests, thus obtaining an important protein supplement to their diet. This strategy helped to restore the tree duck population to its former extent. More important, more than 80 local people worked directly and indirectly for the project and a number of these put up their own nesting boxes. This in turn contributed to a substantial appreciation by the local communities of the importance of Laguna El Jocotal and the sustainable benefits that it can provide.

▼ *The Gandoca-Manzanillo Wildlife Refuge Ramsar Site, Costa Rica. This small site protects an important diversity of freshwater and marine wetlands, including mangroves, freshwater and brackish marshes, a coastal lagoon, and coral reefs.*

Multiple Uses of Mangroves

Along the Pacific coast of Central America as well as in the Caribbean, cutting mangroves to burn and produce charcoal is an important subsistence activity; charcoal is not only used as a home cooking fuel by local people but is also sold to urban centers and tourist resorts. To make charcoal, the cut trunks and branches are stacked in a mound and covered with mud or soil and vegetation to ensure that the wood smolders slowly for several days. The longer the wood smolders the better the quality of the charcoal.

In Central America, it is common for the larger trees to be used for construction purposes in coastal areas, while thinner poles and branches form an important source of roofing materials in many coastal areas. In Nicaragua bark is harvested for the extraction of tannins used in curing leather.

In many areas all of these multiple uses are possible, together with fishery and wildlife harvest. However, this requires careful management in order to avoid overuse of one or more mangrove product. Charcoal cutting in particular can severely degrade mangrove forests, especially if *Rhizophora* is used, as this species does not coppice like *Avicennia* and *Laguncularia*. Where mangrove stands are cleared entirely, the exposed soil dries out quickly and may become too salty for seedlings to grow. Good management preserves mature trees at intervals to provide shade and to aid re-seeding and may involve selective replanting.

Biodiversity

On the Caribbean coast of Costa Rica, Tortuguero National Park alone supports 405 species of bird, 184 fish and 97 mammals. The mammals include a number of endangered species, such as the manatee (*Trichechus manatus*), the jaguar (*Felis onca*), the puma (*Felis concolor*), the tapir (*Tapirus terrestris*) and four species of monkey. Resident and migratory birds are also numerous, among them are found great white herons (*Casmerodius albus*), wood storks (*Mycteria americana*), white ibis (*Eudocimus albus*) and snowy egrets (*Egretta thula*).

▼ *Tapir (Tapirus terrestris)*

As well as the numerous fish, the rich aquatic fauna includes several species of freshwater turtle, such as *Chrysemis ornata*, and sea turtle

(*Chelonia mydas* and *Dermochelis coriacea*); and crocodilians such as *Caiman crocodilus* and *Crocodylus acutus*. These populations were once heavily depleted but have now recovered substantially, particularly after the establishment of the neighboring Caño Negro Wildlife Refuge in 1984. However poaching remains a significant concern.

The wetlands of this border region between Costa Rica and Nicaragua are particularly important for the fish populations that they support. Some, such as the tarpon (*Megalops atlanticus*), provide a major attraction for sport fishing, while others such as the relict fish *Atractosteus tropicus* and the freshwater bullsharks (*Cacharhinus leucas*) of the Lake of Nicaragua are of special interest to the scientific community. Different species of cichlid fish (*Cichlasoma* spp.) and *Rhamdia guatemalensis* are the base of subsistence fisheries.

As more and more of the freshwater wetlands on the Pacific coast of Nicaragua and Costa Rica are being affected by drainage and water diversion, the biological role of the remaining wetlands in this border area has become increasingly significant. In recognition of this, Costa Rica has listed Caño Negro, Tortuguero National Park, and Barra del Colorado Wildlife Refuge as Ramsar sites.

▲ *Snowy egret* (Egretta thula)

Northern South America and the Amazon Basin

The northern half of South America is dominated by the Amazon drainage basin. Covering a total of some 2.7 million square miles (7 million square kilometers) the basin contains some of the most extensive floodplain forests, known as *várzea* forest, in the world. The rivers which unite to form the River Amazon flow from headwaters in Bolivia, Peru, Ecuador, Venezuela and Brazil. The largest is the Solimoes, which, after joining the Río Negro east of Manaus, northern Brazil, forms the middle Amazon and flows over 720 miles (1,200 kilometers) to the sea. At its mouth, near Belém, on the north coast of Brazil, the Amazon releases over a sixth of the freshwater carried by all the world's rivers.

The mighty Amazon

Along its course, the Amazon, and its tributaries, inundate 19,000–23,000 square miles (50,000–60,000 square kilometers) of the middle and lower Amazon for several months each year. The deepest parts of this vast forested floodplain are occupied by lagoons which expand and shrink with the flood. The most extensive lagoons cover over 770 square miles (2,000 square kilometers). In the nutrient-poor rivers there are numerous palm-shrouded channels, or *igapos*. At the mouth of the river, the tidal influence extends as far as 600 miles (1,000 kilometers) inland, creating vast areas of seasonal and tidal floodplain. The biggest island is Marajo, which divides the estuary into two branches. The estuary includes areas of flooded grassland where only the *tesos*, mounds built by pre-Columbian Indians, emerge above the water.

Moving east of the Amazon, the coast of Para and Marañhao is indented with more than 35 major inlets and estuaries, each fringed with mangrove swamps. Inland, there are fresh to brackish lagoons and marshes, riverine marshes, areas of seasonally flooded grassland, palm groves and patches of forest.

North of the Amazon Basin lies the Orinoco River, which rises on the Guyana Shield. Here the river shares several tributaries with the Río Negro, the most famous being the Casiquiare. Northeast of this region, the Caroni River rises in the Gran Sabana, joining the Orinoco some 120 miles (200 kilometers) before it reaches the sea.

In the highlands of Colombia and Ecuador, along the upper courses of the Río Magdalena, there are bogs,

Santuario Nacional Los Manglares de Tumbes

ANDES PE

PACIFIC
OCEAN   Lima•
Zona Reservada Los
Pantanos de Villa

N
↑

| 0 km | 500 |
| 0 miles | 250 |

■ Ramsar Sites

◆ Ramsar Sites/
   Parks and Reserves

□ Water Bodies

□ Freshwater Marsh/
   Floodplains

□ Mangroves

□ Swamp/Flooded Forest

Key to numbered Ramsar sites
COLOMBIA
1 Laguna de la Cocha

ECUADOR
2 Reserva Ecológica Cayapas-Mataje
3 Laguna de Cube
4 Reserva Biológica Limoncocha
5 La Segua
6 Machalilla
7 Abras de Mantequilla
8 Isla Santay
9 Manglares Churute
10 Refugio de Vida Silvestre Isla Santa Clara
11 Parque Nacional Cajas

marshes and oxbow lakes which are important for resident and migratory waterfowl. And in the High Andes of Peru there are many glacial lakes and mountain rivers, many of which are particularly important habitats for waterfowl species, some of which are endemic to the area.

105

Coastal wetlands

On the Caribbean coast of Colombia, the Ciénaga Grande de Santa Marta forms a complex of wetland habitats near the mouth of the Río Magdalena. Near the coast there are large shallow, brackish to saline lagoons and mangrove swamps, while further inland extensive, freshwater lakes, marshes and swamp forest are regularly flooded by the Río Magdalena. Covering a total area of 190 square miles (500 square kilometers), this complex of wetlands is the most important on the Caribbean coast of Colombia, and provides an important fishery for oysters and shrimps, sport and subsistence hunting and harvest of mangroves. Waterfowl are abundant, with large numbers of resident breeding species and migrants from North America.

Given the low rainfall of the Caribbean coast of Colombia, the Ciénaga Grande is dependent on the inflow of freshwater provided by the melting snows of the Sierra Nevada de Santa Marta. Canalization of the Magdalena, however, has reduced water flow to the lagoon. As a result the salinity of the lagoon has increased and caused significant mangrove loss. In addition when the road connecting the coastal cities of Barranquilla and Santa Marta was built, the free flow of water between the lagoon and the sea was stopped, so further degrading the mangroves of the Ciénaga Grande. To address this problem culverts have been placed at intervals along the road and sea water now enters the lagoon through these. However, although areas near the culverts are now in better condition, the benefits are localized. Furthermore, the evaporation of seawater is exacerbating the salinization problems, and experts believe that the Ciénaga Grande will recover fully only when freshwater flow from the Río Magdalena is once again established.

The Llanos

In a belt running from the Cordillera de Merida in the west of Venezuela to the Guyana Shield and the Orinoco River to the south and east, lie the Venezuelan Llanos, part of the

Baranquilla

Sistema Delta Estuarino del Río Magdalena, Ciénaga Grande Sta.Marta

Panama City

Bahía de Panamá

Punta Patiño

PANAMA

PACIFIC OCEAN

Delta del Río Baudó

San Juan River Delta

Reserva Ecológica Cayapas-Mataje

Laguna de la Cocha

ECUADOR

Cauca

Magdalena

ANDES MOUNTAINS

Bogotá

floodplain of the Apure and Orinoco rivers and their tributaries. The Llanos are among the largest wetland areas in South America, covering over 40,000 square miles (100,000 square kilometers) in Venezuela alone (they also extend across the border into Colombia). This vast wetland region forms a mosaic of slow-flowing rivers and streams, oxbow lakes, riverine marshes and swamp forest, permanent and seasonal freshwater lakes, ponds and marshes, and large areas of seasonally inundated grassland and palm savanna. For six months of the year, the Llanos become an ocean of water and grass, turning, as the floodwaters recede, to a dry, dusty plain.

The cycle of flood and drought which characterizes the Llanos determines the seasonal cycle of the fauna and flora of the floodplain. The summer dry season is a time of reduced growth, when annual plants die off and the smoke of frequent forest fires drifts across the dry savanna. During this time, most animals move to the valley bottoms and along the rivers, where the land remains wet. With the winter rains comes the flood which fills the rivers and lakes, and stretches across the plain. This stimulates the germination of annual plants and the sprouting of perennials, nourishing the region's food chains.

Ramsar Sites

Ramsar Sites/
Parks and Reserves

Water Bodies

Freshwater Marsh/
Floodplains

Mangroves

Swamp/Flooded Forest

▽ Delta

○ Dam

▲ *Young caiman*
(Caiman crocodilus)

A wealth of life

The floodplain supports a rich and varied flora and fauna. The vast expanses of aquatic grasses are dotted with palms (*Mauritia* spp. and *Copernicia tectorum*), which are used by the locals for roofing material. In general, however, the rigors of the drought and fires limit the development of more complex plant communities. Instead, well-developed woodlands, for example, tend to establish along river beds, where they are protected from fire.

The Llanos' rivers are rich in fish, many of which are caught and eaten. Other fish, such as the stingray (*Dasyatis americana*) and the electric eel (*Electrophorus electricus*), can cause painful wounds and are feared by the local population. Most famous of all the fish, however, is the caribe or red piranha (*Serrasalmus nattereri*), which has become world famous. When local people have to swim across rivers infested with piranhas, they fear them much more than the local crocodilian, the caiman.

▼ *Capybara*
(Hydrochoerus
hydrochaeris) with baby
on the Llanos of
Venezuela.

The caiman (*Caiman crocodilus*) is one of the most important residents of the Llanos. Once described as being so numerous as to form a "moving pavement" in

▲ *Green tree frog*
(Hylidae cinerea)

the water, poaching has reduced numbers substantially. Fortunately, conservation efforts are increasing and the outlook is now much more positive.

The Llanos also provide one of the world's most important habitats for wading birds. Over 100 breeding colonies of herons, storks and ibises have been recorded in a single year, with the largest holding 32,000 pairs. In 1983, 65,000 pairs of scarlet ibis (*Eudocimus ruber*) were located in 22 colonies, as well as over 5,500 wood storks (*Mycteria americana*) and 185 jabiru (*Jabiru mycteria*).

The grasslands of the Llanos support a rich variety of herbivores, notably the capybara (*Hydrochoerus hydrochaeris*) and the white-tailed deer (*Odocoileus virginianus*). Today, the deer mix with the cattle, which combined total 10 million individuals grazing on the floodplain, and together they have come to characterize the Llanos in the minds of many people. At the top of the food chain is the jaguar (*Panthera onca*), which is affectionately known as "tio tigre" (uncle tiger) by the local people.

Peoples of the Llanos

Prior to European colonization of the Americas, the Llanos were inhabited by different tribes over the centuries, stretching as far back as 15,000 years ago. These tribes were hunter-gatherers who lived off the rich resources of fish, wildlife and wild plants. Today, less than 10,000 indigenous people remain.

European settlement of the Llanos did not begin until relatively recently. The first explorers saw it as empty, without the gold and other easy riches that they sought. Only in the late 1700s and early 1800s did efforts at colonization lead to the beginnings of a long-term presence based upon cattle and centered around missions. Even today the Llanos have a low population density.

The past 200 years have left their mark on the Llanos. The most significant change is the progressive impoverishment of the region's soils under continuous grazing, the loss of several species through fires, and the slow extermination of others through illegal hunting. However, there is today a substantial increase in conservation investment in the Llanos, with government and non governmental organizations, and private landowners, working together to establish forms of land use that can yield economic benefits without destroying the natural riches of the region.

The man-made wetlands of Guyana

The coast of Guyana is bordered by a plain about 15 to 21 miles (25 to 35 kilometers) wide, most of which lies below the level of mean high tide. For over three centuries this coastal plain has been the focus of development investment. As early as 1621, the Dutch West India Company established the Essequibo colony. Between 1655 and 1680 a sea wall was built to reclaim land from both the sea and the freshwater swamps. An interlaced network of drainage channels was constructed, and at the seaward end of these channels kokers were installed, gatelike devices that allow the land to be drained at low tide. Plantations were laid out in strips running inland, and a dam and a network of irrigation canals were built to provide water for farmlands and populated areas.

▼ *The giant aquarium. The remarkable diversity of the Amazon waters – rapid in some places, sluggish elsewhere, clear in parts, opaque with debris in others – breeds an unequaled variety of fish life. One of the largest families, the characids, displays a wide range of hues and habits, from the neon tetra* Hyphessobrycon innesi *to the savage piranha* Serrasalmus spp. *which, in a school, has been reported to strip the*

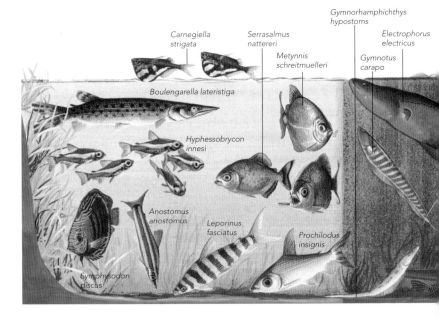

Gymnorhamphichthys hypostoms

Carnegiella strigata

Serrasalmus nattereri

Metynnis schreitmuelleri

Electrophorus electricus

Gymnotus carapo

Boulengarella lateristiga

Hyphessobrycon innesi

Anostomus anostomus

Leporinus fasciatus

Prochilodus insignis

Symphysodon discus

Today, more than 90 percent of Guyana's population lives on the coastal plain, and investment in drainage for agriculture and other development projects has continued. Over the past 40 years, the sea defense, and drainage and irrigation system have been expanded through several major land development projects.

flesh from a 100lb (45kg) capybara in under a minute. The hatchet fish Carnegiella spp. *are true "fliers" in pursuit of airborne insects. The electric gymnotids (center) are nocturnal feeders and include the knifefish* Gymnotus carapo *and the electric eel* Electrophorus electricus *which stuns its victims with a shock of up to 600 volts. The arapaima is the largest freshwater fish in the world – reaching 200lb (90kg).*

The cost of agriculture

The net result of three centuries of modification of the coastal plain is that the region today forms one of the largest man-made wetlands in the world. More than 770 square miles (2,000 square kilometers) of land within the coastal strip are irrigated and drained by 8,400 miles (14,000 kilometers) of man-made channels. Landward, the artificial wetlands and their associated swamps cover 2,100 square miles (5,400 square kilometers). However, in common with the rest of the world, Guyana's wildlife has suffered from the conflict between agriculture and wildlife conservation. Misuse of pesticides and herbicides is believed to have killed large numbers of fish, hatchling caiman and other aquatic wildlife, and spectacled caiman populations have been depleted through commercial hunting.

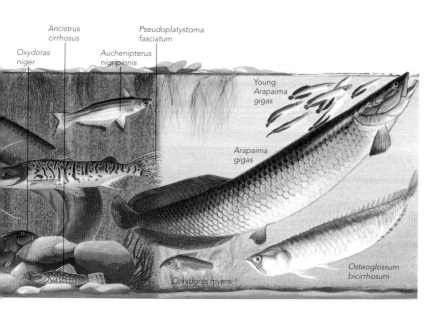

Oxydoras niger

Ancistrus cirrhosus

Auchenipterus nigripinnis

Pseudoplatystoma fasciatum

Young Arapaima gigas

Arapaima gigas

Corydoras myersi

Osteoglossum bicirrhosum

The Amazon

Since its encounter with the first Europeans in 1542, the River Amazon has fascinated successive generations of modern society. Coming only 10 years after the Incas of western Peru had been overwhelmed by Francisco Pizarro, the discovery of this mighty river led to successive expeditions in search of further riches. The Europeans knew of the societies which had flourished along the Nile, the Tigris-Euphrates, the Ganges and the Indus, and sought similar wealth along this, the largest of the world's rivers. That they found only Indian tribes living from the resources of the forest and river led to disillusionment, and when in 1639 Pedro Teixeira traveled 2,000 miles upstream and claimed for Portugal all of the land which now lies east of Ecuador, Spain saw this as being of little importance.

Over three centuries later, the Amazon Basin remains one of the world's great wilderness; a river of awesome power bordered by the Earth's largest

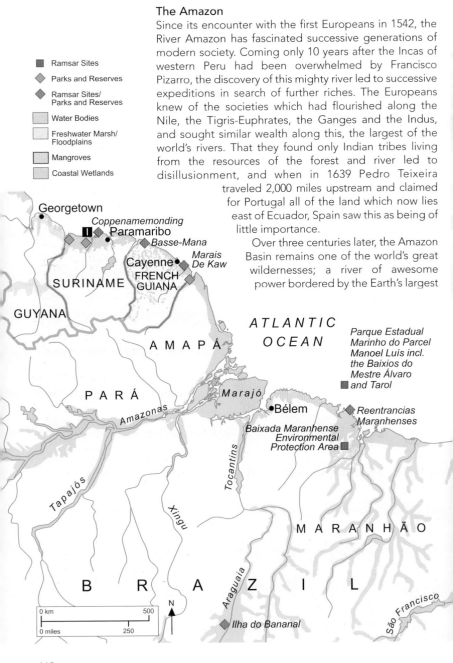

Ramsar Sites
Parks and Reserves
Ramsar Sites/
Parks and Reserves
Water Bodies
Freshwater Marsh/
Floodplains
Mangroves
Coastal Wetlands

Georgetown
Coppenamemonding
Paramaribo
Basse-Mana
Marais
De Kaw
Cayenne
FRENCH
SURINAME GUIANA
GUYANA

ATLANTIC
OCEAN

Parque Estadual
Marinho do Parcel
Manoel Luís incl.
the Baixios do
Mestre Álvaro
and Tarol

AMAPÁ

PARÁ

Amazonas

Marajó

Bélem

Reentrancias
Maranhenses

Baixada Maranhense
Environmental
Protection Area

Tocantins

Tapajós

Xingu

MARANHÃO

BRAZIL

Araguaia

São Francisco

0 km   500
0 miles   250

N

Ilha do Bananal

▲ *Amazon community. Along the Amazon's long journey to the sea it supports many thousands of communities dotted along its banks. Many live in temporary shelters, while others live in houses built on stilts as an adaptation to the rivers annual flood.*

expanse of tropical forest. Today, the Amazon is South America's new frontier, an area rich in mineral resources where many hope to find solutions to problems of wide-spread poverty, rising population, inflation and external debt. As the forest has fallen, concern for the future of the Amazon region and the people who live there has received growing international concern, giving rise to claims of foreign interference in national affairs. To date, most of the debate over the Amazon has focused upon the forest resources with little consideration given to the river and its associated lakes and floodplains, arguably the most productive but fragile parts of the system.

▮ Coppename River Mouth

The wide, tidal mud flats, lagoons and brackish herbaceous swamps in the estuary of the Coppename and Suriname rivers are typical of the coastal wetlands of the Guyanas and northern Brazil. Protected since 1953 this 46-sq mile (120-sq km) reserve was established as a Ramsar Site in 1985 and in 1989 as a Western Hemisphere Shorebird Reserve. Up to 750,00 Semi-palmated sandpipers (Calidris pusilla) have been recorded using the reserve, together with tens of thousands of other migratory shorebirds.

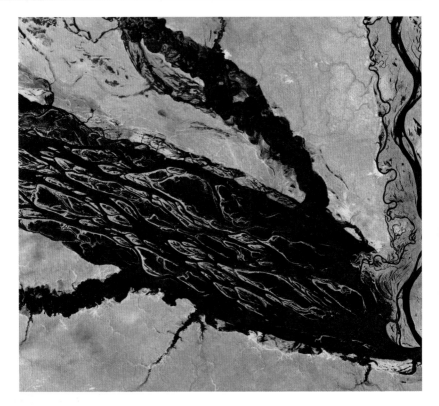

A stretched resource

Today, fishing is the most important economic activity along the rivers and in the adjacent *várzea* floodplain forest and lakes. Because terrestrial sources of food such as cassava provide little protein, river communities have relied heavily on fishing to provide essential dietary protein. For example, 61 percent of the animal protein eaten by rural people in the Ucayali Valley of Peru comes from fish, while in Brazil fish provide over 60 percent of protein consumed by the 100,000 families that live along the river banks in the States of Amazonas, Para and Amapa.

Traditional small-scale fishing on the Amazon is however now in conflict with larger-scale fishing from the ports of Manaus and Belém. As the rural poor have migrated to these urban centers in search of work, demand for fish protein has grown substantially. Commercial fishing has intensified in response to this demand, and also to meet the needs of the larger towns

▲ *Jau National Park, Brazil. Fed by multiple waterways, Brazil's Rio Negro is the Amazon's largest tributary. The mosaic of partially submerged islands visible in the channel disappears when rainy season downpours raise the water level. This image was acquired by Landsat 7's Enhanced Thematic Mapper plus (ETM+) sensor.*

Amazonas

Marajó

Tocantins

•Bélem

▦ *Parque Estadual Marinho do Parcel Manoel Luís*
incl. the Baixios do Mestre Alvaro and Tarol

◈ *Reentrancias Maranhenses*

▦ *Baixada Maranhense*
Environmental Protection Area

P A R Á   M A R A N H Ã O

B   R   A   Z   I   L   Recife•

São Francisco

◈*Ilha do Bananal*

•Brasília

ATLANTIC

OCEAN

Paranaíba

Rio de Janeiro •

	Ramsar Sites
	Ramsar Sites/ Parks and Reserves
	Water Bodies
	Freshwater Marsh/ Floodplains
	Mangroves
	Coastal Wetlands
	Swamp/Flooded Forest

| 0 km | 500 |
| 0 miles | 250 |

N ▲

of the south. In turn this increased commercial fishing has resulted in overfishing and this has exacerbated the problems of the rural communities.

In an effort to protect the floodplain fisheries and the people dependent upon them, a growing number of initiatives to establish fishery reserves in the floodplain lakes are being pursued. Community leaders urging the establishment of these reserves, however, have had to confront powerful commercial interests currently overfishing the systems, and several leaders have been killed.

Consequences and action

The development boom in the Amazon Basin has led to widespread deforestation, resulting in soil erosion, river siltation and declining productivity of many of the aquatic ecosystems. Meanwhile, the expansion of gold mining in many of the southern tributaries has resulted in severe

◄ *Aerial view of the Amazon river in flood. Covering 4.26 sq mi (6.86 million sq km) the Amazon's drainage basin is the world's largest and covers 25% of the surface of South America. The water carried by the river represents one fifth of the volume carried by all of the world's rivers. The river's mouth is 250 km wide with the island of marajo, the largest river island in the world, lying between the northern and southern banks. Marajo covers an area the size of Switzerland and large parts lie underwater during the flood season.*

mercury contamination. Many areas of várzea have also been cleared for livestock and commercial crops such as jute, corn and rice. While it is hoped that such agricultural practices will benefit from the fertile soils of the floodplain and generate sustainable high yields, it is at the expense of traditional river communities, who must find other land or move to the city. Clearance of the floodplain is believed to have led to a 23 percent decline in Amazon fish catches between 1970 and 1975 alone.

Today, international attention is being focused upon the dams of the Amazon Basin. So far only seven are in operation in the Brazilian basin, but there are others in Ecuador and Venezuela, and across the basin a total of 384 feasible sites for dam construction have been identified. While increased supplies of clean power are essential for effective and environmentally sound exploitation of the region's mineral wealth, the completed dams have been plagued with cost overruns, environmental and social impacts and doubts over their long-term viability.

The development pressures upon the Amazon are enormous. A substantial part of the basin's natural wetlands have already been destroyed by human activity. However, it is not too late for action to address these. Legislation and effective policing procedures are needed to stop mercury pollution from gold mining; proposals for new dams must consider the total impact on the region and its people, and systems of integrated resource use on the floodplain need to be developed and promoted as part of government policy.

Southern South America

From the high-altitude lakes of the Andes, through the vast floodplains of the Paraná–Paraguay river basin to the mangroves and lagoons of the Brazilian coast, the southern cone of South America supports some of the most varied wetlands in the world. Although these wetlands have supported human populations for thousands of years, today the southern cone is the site of growing conflict over conservation and development policy. The

1 Rio Grande do Sul

The coastal plain of Rio Grande do Sul in southern Brazil supports a diverse assemblage of lakes and lagoons of various sizes which lie behind a chain of barrier islands and dune systems. The largest lagoon in Brazil is the Lagoa dos Patos, which stretches 150 miles (250 km) and covers 4,000 sq miles (10,360 sq km). Increasing population and industrial development have increased pressure upon the coastal areas of Río Grande do Sul and industrial and domestic effluents have degraded the lagoons. In the surrounding marshes drainage for pastureland and rice cultivation has given rise to calls for reserves to be established.

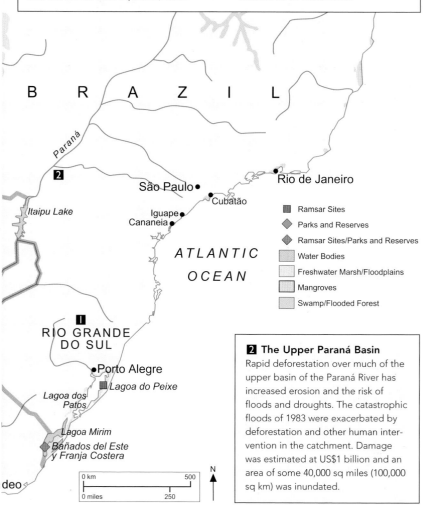

2 The Upper Paraná Basin

Rapid deforestation over much of the upper basin of the Paraná River has increased erosion and the risk of floods and droughts. The catastrophic floods of 1983 were exacerbated by deforestation and other human intervention in the catchment. Damage was estimated at US$1 billion and an area of some 40,000 sq miles (100,000 sq km) was inundated.

nations of the region are seeking to maximize economic output and raise GNP in the face of rising population and external debt. In support of this, increased investment in wetland conservation is needed in an effort to ensure that wetlands and their benefits are maintained.

The Paraná

Stretching 2,500 miles (4,000 kilometers) from the interior of Brazil to its mouth at Buenos Aires, the Paraná River is the second largest in South America; its basin drains 1.1 million square miles (2.8 million square kilometers). Just north of the Argentine cities of Resistencia and Corrientes on the border with Paraguay, the Paraná is joined by the Paraguay River, which also flows through Brazil for 1,350 miles (2,250 kilometers). The vast region surrounding these river courses is predominantly flat, has a benign climate and possesses some of the most fertile soils on the continent, factors which have combined to make this the most densely populated and industrialized region of South America. The two largest cities of South America, Sâo Paulo and Buenos Aires, lie within the Paraná-Paraguay Basin.

Matching this economic importance, the floodplains of the basin are among the most important in the world. For most of the year, the Pantanal covers over 54,000 square miles (140,000 square kilometers), abutting the borders of Brazil, Paraguay and Bolivia. During the flood season, the Pantanal spreads out to flood 96,500 square miles (250,000 square kilometers) – over 6 times the size of Belgium. For this reason the Pantanal is widely regarded as the largest wetland in the world. To the south, the basin's floodplains continue in virtually uninterrupted progression. They spread through the wet prairielike Chaco region of Paraguay and Argentina, join the floodplain of the Paraná River south of the confluence with the Paraguay, and terminate in the delta of the Paraná River to the northwest of Buenos Aires.

The annual flood cycle of the river is critical for the livestock industry which has developed along the river. During the periods of receding flood, small lakes and areas of extensive pasture remain, providing a rich food source for large herds of cattle. During the flood the cattle are removed to higher land and the floodplains are then the domaine of the fish which migrate upstream to breed. Young fish normally colonize the lakes of the floodplain, where they remain for one or two years before returning to the river with the flood.

▲ Native llamas share the Titicaca grasslands with introduced sheep. The domesticated llama (Llama guanicoe glama) is larger than the wild guanaco (Llama guanicoe) from which it was bred thousands of years ago. Although llamas are now raised mainly for wool, they were once important beasts of burden, prized for being able to carry heavy loads.

Bañados del Este

The Bañados del Este, the coastal lagoons and freshwater marshes on the shores of Laguna Merin, cover 770 square miles (2,000 square kilometers) and are among the most important natural ecosystems in Uruguay. In 1984, the Bañados were listed under the Ramsar Convention when Uruguay signed the treaty. They support a high vertebrate and invertebrate diversity, including 35 percent of the freshwater fish species, 47 percent of amphibians, 58 percent of reptiles, 42 percent of birds and 51 percent of the mammals of Uruguay. One of the characteristic landscapes of the Bañados del Este are the groves of butia palm, which cover a total of 270 square miles (700 square kilometers). These are now threatened by cattle whose grazing prevents the growth of young trees.

Wetlands of the Puna

The cold, arid Andean altiplano, known as the Puna, is characterized by extensive tablelands that lie between 10,500 and 14,800 feet (3,200 and 4,500 meters) above sea level. Here there is a diversity of lakes of various sizes and salinities, together with temporary lagoons, freshwater marshes known as bofedales, and bogs. Of the lakes, Lake Titicaca is the most famous. According to legend

Humedal el Yali
Santiago

PACIFIC
OCEAN

URUGUAY

Montevideo
Buenos Aires

Laguna de Llancanelo

Río de
la Plata

Bahía de Samborombón

A R G E N T I N A

CHILE

Bahía Blanca

Colorado

Laguna Blanca

Carlos Anwandter Sanctuary

Negro

Puerto Montt

Ramsar Sites

Parks and Reserves

Ramsar Sites/
Parks and Reserves

Water Bodies

Freshwater Marsh/
Floodplains

Peatlands

A
N
D
E
S

Chubut

ATLANTIC
OCEAN

Taitao
Peninsula

Deseado

Falkland
Islands

Bertha's Beach

Sea Lion Island

Magellan's Strait

Reserva Costa Atlantica
de Tierra del Fuego

Tierra
del
Fuego

N

0 km   500

0 miles   250

Cape Horn

the sun gods created Manco Capac and his wife Mama Ocllo, the first of the Incas, on Lake Titicaca's Isla del Sol (Island of the Sun). From there they spread out to establish the Inca empire.

The highest navigable lake in the world and the largest in South America, Lake Titicaca was held sacred by the Incas. Prior to the arrival of the Europeans, the lake supported a healthy Inca economy based principally upon rearing llamas and alpacas, as well as fishing and commerce. Today, the lake is shared between Peru and Bolivia, and a sizable population continues to live on its shores and depend upon its fringing wetlands.

The domestic herds are now composed of vicuñas and cattle. The submerged aquatic vegetation, "yacco", is collected as cattle feed. In addition, reeds (*Scirpus* sp.) are used in handicrafts and as food. An estimated 5,000–6,000 tons of fish are caught annually for direct consumption and sale.

Icy lakes

Although they lie at high altitude and share a cold alpine climate,the lakes of the Puna are home to many endemic species and important communities of waterbirds. Lake Junín in Peru, with endemic species of grebe (*Podiceps taczanowskii*) and frog (*Batracophrynus macrostomus*), is one of the most famous and best studied of these lakes. Three species of flamingo live in the Puna, two of which are endemic to the region. Of these the Andean flamingo (*phoenicoparrus andinus*) is the biggest and rarest, and Lake Atacama in Chile is its only permanent breeding site. James' flamingo (*Phoenicoparrus jamesi*) is the smallest and nests mainly in the Laguna Colorado in Bolivia, while the third species, the Chilean flamingo (*Phoenicopterus chilensis*), is more widely distributed throughout the south of the continent. It is thought that the flamingos, which feed by filtering phytoplankton and zooplankton, are confined to salt lakes where there are no fish. The most likely reason for this is that fish consume the greater part of the available food supply in other lakes.

Despite the relative remoteness of the Puna's wetlands they face a diversity of pressures. In Lake Titicaca the introduction of trout has resulted in a decline in native species, while indiscriminate burning of the reed beds, and hunting and grazing, all threaten to degrade the ecosystem upon which both people and wildlife depend. Many of the region's lakes are contaminated by run-off from the many precious-metal mines. Lake Junín, for example, has been contaminated by residues from the lead, copper and zinc mines of the Cerro Pasco-La Oroya.

Conservation strategies
Despite the problems they face, important progress has been made in the past decade or so to protect the Puna's wetlands. A concentrated effort by national and international conservation organizations halted plans to use water from Lake Chungara in Chile for irrigation; parts of lakes Titicaca and Junín are now protected areas, Laguna Colorada in Bolivia is listed under the Ramsar Convention, and the Laguna de Pozuelos in Argentina, which in 1992 was listed as the country's first Ramsar Site, is now a Biosphere Reserve.

The Pantanal
Seen from the air, the Pantanal resembles a patchwork quilt; floodplain lakes and marshes interspersed with one of the most diverse tree floras on the continent. Areas of higher ground provide important refuges for many species of birds and mammals as well as serving as nesting sites for caiman, and greatly enhance the richness of the flora and fauna of this vast wetland.

▼ Royal Water-Lilies in the Pantanal. Victoria Regia water lily leaves with a flower just opened, early in the morning. The flower turns from white to pink during the day.

▲ The jaguar is the largest and most powerful of the American cats, and needs no special adaptation to take advantage of wetland habitats. Terrestrial mammals are the jaguar's main prey, although they frequently take caiman from the water.

On the ground, the Pantanal supports some of the largest concentrations of waterbirds in Latin America and, together with the Llanos of Venezuela, provides a critical stronghold for the Jabiru stork (*Jabiru mycteria*), the largest of the region's wading birds. The density and diversity of mammals is, however, limited by the incidence of extreme floods. Five times in the 1900s – in 1905, 1920, 1932, 1959 and 1974 – exceptionally high floods inundated large parts of the Pantanal. Small terrestrial mammals are therefore largely absent from the flooded areas, except for those which managed to find refuge on high ground. Intensive cattle management has also converted large areas of the forest habitat to pasture, thus changing a complex habitat to a simple one, and at the same time eliminating most forest mammals.

Home of the jaguar

Although the Pantanal's animal population may be limited in its diversity, it is one of the most spectacular in Latin America. Despite continued poaching, the Pantanal is one of the remaining strongholds for the jaguar, with a population estimated at 3,500 in 1986. Each adult male ranges over an area of 30–50 square miles (50–80 square kilometers) with females occupying half of this. Both males and females feed on a wide variety of animals from coati (*Nasua nasua*) and caiman (*Caiman yacare*) to tapir (*Tapirus terrestris*). Cattle are an important prey, although kills account for only a small percentage of the animals dying annually. In the Pocone district, which contains the largest remaining jaguar population in the Pantanal, drowning, disease and starvation reduced the cattle population from 700,000 before floods in 1974 to 180,000 in the mid-1980s.

The main wild prey of the jaguar is the capybara (*Hydrochoerus hydrochaeris*), although populations of this large rodent were also reduced severely following the 1974 flood. Deprived of their favored habitat on the floodplain, the capybara crowded onto the higher ground, where they were vulnerable to diseases. The marsh deer (*Blastocerus dichotomus*), the largest deer in South America, was also affected by the flood.

Skin trade

The Pantanal resident that has suffered most from poaching is the caiman. Groups of professional hunters

▼ *Pantanal. The Pantanal forms a unique mosaic of wetlands, grasslands and forests that support a rich diversity of wildlife. Many species congregate in large numbers, especially during the dry season when shallow lakes provide rich feeding habitats.*

used the area during the dry season and operated on private lands with or without the landowners' permission. In the 1980s an estimated one million hides were smuggled out of Brazil each year, being taken directly, or indirectly through Bolivia, to Asunción in Paraguay. In the face of this pressure, populations declined dramatically in most areas, but the species is resilient and is still common in many areas. They have survived because, as well as being extremely secretive and wary, they have the ability to reproduce at a young age and adapt to different habitat types. As conservation investments in the Pantanal increase there is good reason to hope that the future of the caiman and the other wildlife of this unique wetland will be secure.

The Western Hemisphere Shorebird Reserve Network

Many shorebirds undertake extraordinarily long migrations, some from Alaska and northern Canada to Tierra del Fuego, Argentina. Along the way they rest and feed at a few widely separated and utterly crucial sites. The Western Hemisphere Shorebird Reserve Network (WHSRN) was established to help protect these sites and the places where the birds spend their nonbreeding seasons.

▲ *Laguna de Mulas Muertas, Laguna Brava Ramsar site in Argentina. Laguna Brava is a network of saline and hyper-saline shallow lagoons which include characteristic high Andean marshy meadows (bofedales) that occur above 9,000 ft (3,000 m). The reserve supports a rich bird population including James and Andean flamingoes (Phoenicoparrus jamesi and P. andinus) during the summer breeding season.*

**Western Hemisphere
Shorebird Reserve
Network**

● Sites of Hemispheric
Importance

● Sites of International
Importance

● Sites of Regional
Importance

WHSRN is a voluntary, nonregulatory coalition of governments, nongovernmental organizations, and private companies launched in 1985 through the efforts of the Manomet Bird Observatory (now Manomet Center for Conservation Sciences), the Canadian Wildlife Service and the Philadelphia Academy of Natural Sciences. Today it helps protect 60 sites (see map) totaling 20 million acres in eight countries, with more than 250 partners in the private, governmental and nongovernmental sectors.

Hidrovia

The Paraná–Paraguay river system has, for hundreds of years, provided an important transportation artery into the heart of the South American continent. However as plans for regional economic integration have developed over the past few decades, growing attention has been focused on whether river navigation can be improved. This has led to plans to alter the river so that it is navigable between Caceres in Brazil and the Port of Nueva Palmira in Uruguay. It is thought that such an alteration would allow the cheap transportation of produce from the interior. However, while there is widespread enthusiasm for the project, which could do much for the commercial development of the region, there is also concern over its possible impact upon the natural systems of the basin, and over the economic costs of such changes. The most serious threat is to the Pantanal, where there are plans to dredge the River Paraguay for 400 miles (670 kilometers) between Corumba and Nuevo Caceres. This would not only result in major changes in the seasonal flooding cycle of the Pantanal and affect the region's rich animal and plant life, but it would also alter the pattern of flooding downstream. The Pantanal currently absorbs flood waters and releases these slowly during the course of the dry season. As a result, there is a difference of about four months between the floods of the Paraguay and Paraná rivers. If the Paraguay is dredged where it passes through the Pantanal, the flow of the river will accelerate and bring the timing of the flood peaks of the two rivers closer together. The effect would be an increase in the overall height of flooding downstream.

Northern Europe

As the ice sheets moved back and forth over northern Europe, they created a heavily modified landscape with eroded hollows, glacial deposits and confused drainage. Today, the eskers (ridges of sand and gravel), drumlins (elongated, rounded mounds), terraces and deltas that are characteristic of Norway, Sweden and Finland are reminders of that glacial past. This post-glacial landscape provides the canvas for the most varied wetland region in Europe.

Peatlands

Bogs and fens cover 23,000 square miles (60,000 square kilometers) in Sweden and Norway, often forming extensive mosaics together with wet forests. In Finland, the original area of bogs was about 42,500 square miles (110,000 square kilometers) – about 30 percent of the land surface. Today, they are thought to cover less than 25,000 square miles (65,000 square kilometers). Their loss is due to natural drying of the peat substrate, drainage for agriculture and mining by the peat industry.

In general, bogs are more common in southern regions, while the largest fens are found in the northern and alpine regions. In summer, these wetlands form important breeding habitats for many northern shorebirds, swans (Cygnus cygnus) and cranes (Grus grus).

For centuries, peatlands have been an important resource in the rural economy. The reindeer herds of the Lapps graze on them in the summer, while substantial harvests of wild fruits are collected, and some, like cloudberry, are harvested on a commercial basis. More recently, interest in the energy potential of peat has increased and in Sweden about 400 sites have

Map Note

When compiling the West and Central European map, the lack of data has made it impossible to ensure that the peatland areas of Scandinavia are shown. A similar lack of data for the remaining floodplains of Europe has also resulted in incomplete coverage for these particular wetland systems.

Key to numbered Ramsar sites

NORWAY
1 Giske Wetlands System
2 Haroya Wetlands System
3 Sandblåst-/Gaustadvågen Nature Reserve
4 Mellandsvågen
5 Havmyran
6 Ørlandet
7 Tautra and Svaet
8 Trondheimfjord Wetland System
9 Øvre Forra
10 Fokstumyra
11 Tufsingdeltaet
12 Kvisleflået
13 Hynna
14 Dokkadelta
15 Åkersvika
16 Mosvasstangen Landscape Protection Area
17 Nordre Tyrifjord
18 Nordre Øyeren
19 Ilene and Presterødkilen
20 Kurefjorden
21 Ora
22 Jaeren
23 Lista Wetlands System

SWEDEN
24 Oldflån-Flån
25 Ånnsjön
26 Tysjöarna
27 Mossaträsk-Stormyran
28 Umeälv delta
29 Sulsjön-Sulån
30 Aloppkölen-Köpmankölen
31 Storkölen
32 Hovran
33 Dalälven-Färnebofjärden
34 Svartån
35 Hjälstaviken
36 Stockholm – outer archipelago
37 Asköviken-Sörfjärden
38 Kvismaren
39 Kilsviken
40 Östen
41 Södra Bråviken
42 Västra Roxen
43 Tåkern
44 Hornborgasjön
45 Dättern
46 Stigfjorden
47 Nordre älv estuary
48 Komosse
49 Dumme mosse
50 Store Mosse and Kävsjön
51 Getterön
52 Träslövsläge - Morups Tånge
53 Tönnersjöheden-Årshultsmyren
54 Fylleån
55 Skälderviken
56 Lundåkra Bay
57 Helgeån
58 Klingavälsån - Krankesjön
59 Falsterbo - Foteviken
60 Mörrumsån-Pukavik Bay
61 Åsnen
62 Blekinge archipelago
63 Emån
64 Öland eastern coastal areas
65 Ottenby
66 Gotland east coast
67 Kallgate-Hejnum
68 Hirsholmene
69 Nordre Rønner
70 Læsø
71 Waters north of Anholt
72 Ulvedybet and Nibe Bredning
73 Randers and Mariager Fjords
74 Vejlerne and Løgstør Bredning
75 Nissum Bredning with Harboøre & Agger Tange
76 Nissum Fjord
77 Stadil and Veststadil Fjords
78 Ringkøbing Fjord

79 Filsø
80 Vadehavet (Wadden Sea)
81 Waters South of Zealand,
    Skælskør Fjord, Glænø
82 Lillebælt
83 Horsens Fjord & Endelave
84 Stavns Fjord
85 Sejrø Bugt, Nekselø Bugt and
    Saltbæk Vig

DENMARK
86 Nærå Coast & Æbelø Fjords
87 South Funen Archipelago
88 Nakskov Fjord & Inner Fjord
89 Waters southeast of Fejo and
    Femo Islands
90 Karrebæk, Dybsø and Avnø
    Fjords

91 Præstø Fjord, Jungshoved
    Nor, Ulvshale and Nyord
92 Maribo Lakes
93 Waters between Lolland and
    Falster
94 Ertholmene

GERMANY
95 Schleswig-Holstein Wadden
    Sea
96 Hamburgisches Wattenmeer
97 Wattenmeer, Elbe - Weser -
    Dreieck
98 Niederelbe, Barnkrug -
    Otterndorf
99 Ostseeboddengewässer
    Westrügen - Hiddensee -
    Zingst

■ Ramsar Sites

◆ Ramsar Sites/
    Parks and Reserves

▭ Water Bodies

▭ Tidal/ Coastal Wetland

▭ Peatland

▽ Deltas

POLAND
100 Slowinski National Park
101 Jezioro Siedmiu Wysp
102 Jezioro Luknajno

131

been accepted for commercial exploitation. In Finland, 4 million tons of peat fuel were produced each year in the 1990s, as well as 300,000 tons of horticultural peat. This destructive use of peat has added to the pressure from agriculture and forestry which has resulted in the drainage of extensive areas over the centuries.

Wet forests

Wet forests, dominated by birch trees (*Betula* spp.), form a transition zone between forest and open wetland habitats. They are key habitats for a large number of threatened or vulnerable animal and plant species, including many insects, mosses and lichens. Birds such as whimbrels (*Numenius phaeopus*) and great snipes (*Gallinago media*) breed there, and they provide a stronghold for mammals such as the European beaver (*Castor fiber*). In Sweden, wet forests and forested bogs cover approximately 19,000 square miles (50,000 square kilometers), and provide a habitat for almost 200 threatened species of flora and fauna. The growing recognition of wet forests as important for other values beside timber production has led to an intensive debate about the impact of modern forestry on natural habitats. As a result, subsidies for drainage operations have largely been withdrawn in Scandinavia, and a stricter legislation on ditching has been imposed on all land users.

▶ *The water level of Lake Hornborga in southwest Sweden was lowered five times in an attempt to gain more arable land. The map dated 1905 shows the lake just before the fourth lowering of the water level. The lake still comprised a large area of water rich in bird life. By 1965, virtually no open water existed, and only sedges and reeds flourished. With initial restoration efforts small areas of open water could be seen in 1988, but it was only after further intensive restoration in the 1990s that the lake's biological value has been restored.*

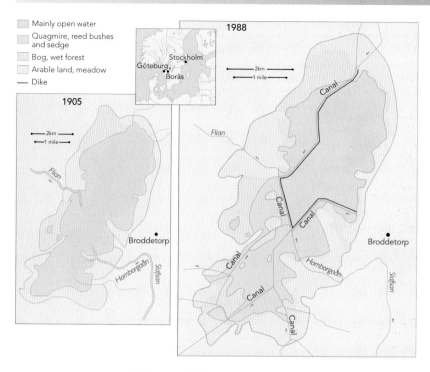

Mainly open water
Quagmire, reed bushes and sedge
Bog, wet forest
Arable land, meadow
— Dike

1905

1988

Stockholm
Göteburg
Borås

2km
1 mile

Flian

Broddetorp

Homborgaån
Slafsan

Flian

Canal

Canal

Canal

Canal

Canal

Canal

Broddetorp

Homborgaån
Slafsan

Rivers and lakes

The Scandinavian peninsula has more than 100,000 lakes, in all covering more than 19,000 square miles (50,000 square kilometers); while in Finland, 10 percent of the total area is covered by freshwater, including 60,000 lakes. Many of the Scandinavian lakes are deep, like Lake Vänern, but the majority are shallow and contain extensive areas of submerged and emergent vegetation. The lakes and watercourses provide critical wildlife habitats and an important recreational resource. Fishing and hunting are especially important.

In the last 150 years, the use of these water resources has become less and less sustainable. A very high proportion of Scandinavian waters has been exploited for agricultural purposes and for power generation. In Sweden, for example, about 75 percent of all suitable lakes and rivers have been regulated as part of hydroelectric developments, bringing irreversible ecological change. A large number of lakes have become totally or partly drained and turned into monotonous reedbeds or shrubby areas, thereby losing their original values. In

◄ *Lake Hammarsjön, one of the nutrient rich lakes included in the Helgean Ramsar site in southern Sweden. These lakes are important for their botanical diversity as well as for the important bird populations they support.*

133

recent years the greatest problems in many areas were caused by acid rain. Borne largely by the prevailing southwesterly winds, rain rich in oxides of sulfur and nitrogen falls upon the region. In Norway, about 40 percent of lakes studied have been recorded to have a pH value of less than 5. In Sweden, more than 17,000 out of a total of 85,000 lakes have become significantly or strongly acidified, and the same general picture is true in Finland. As pH values fall below 5, many plant and animal species can no longer survive, and lakes with pH values of about 4.5 can be entirely empty of fish. Birds and mammals can in turn be seriously affected as their food supply is reduced. Some of these, such as the osprey (*Pandion haliaetus*), have important breeding populations in the region.

Lake Hornborga

Over much of the 1800s and 1900s, Scandinavia and Finland suffered significant loss of wetlands. A programme that involved the lowering of lake water levels to increase the area of arable land was a major cause of loss, with more than a third of lakes affected in some regions of Sweden. One of the most famous of these lakes is Hornborga. By the mid-1900s, Lake Hornborga had been affected by five separate drainage operations. As a result, water level dropped by more than 2 meters (7 feet), and flooding was reduced to a seasonal phenomenon. Large parts of the former lake bed were colonized by bushes, others by extensive reedbeds, while the variety and number of waterfowl using the lake gradually decreased.

The problems were not, however, confined to wildlife. As a result of oxidation and desiccation of the organic soils, ground levels fell and the costs of maintenance of the drainage scheme rose. Farmers began to ask for a reconsideration of the scheme and joined with conservationists who sought to restore the lake for its value as a bird habitat. After a lengthy process of scientific analysis and public debate, a restoration plan was agreed by Government and implemented over the course of the 1990s. This has had a major impact on the lake where many bird species returned. More than 10,000 cranes can

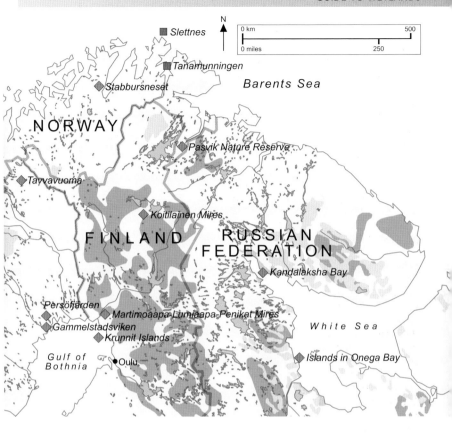

Ramsar Sites

Parks and Reserves

Ramsar Sites/
Parks and Reserves

Water Bodies

Freshwater Marsh/
Floodplains

Tidal/Coastal Wetland

Peatland

now be seen there on a single day during migration in April and all five European grebe species breed at the lake. Some 250,000 people now visit annually.

Denmark

Characterized by its low and gently undulating topography formed by glaciers during the last ice age, Denmark has a very large number of wetlands. Most lie in depressions along rivers, on the fringes of inland freshwater lakes and in coastal areas. Although most of them are small, their combined area and variety support a rich species diversity. In addition, on the edge of the surrounding shallow seas, the coastal fringe supports extensive shallow lagoons and fjords as well as salt marshes.

As is the case in the rest of northern Europe, the main threats to Denmark's wetlands are drainage and cultivation. Over the last 150 years these practices have reduced the country's freshwater wetlands by 75 percent. Meanwhile, the shallow coastal fringe is an easy target for land reclamation, and large coastal wetlands and shallow sea areas have been poldered and cultivated. However in recent years the need to reduce agricultural output has led to the adoption of set-aside policies which provide subsidies to farmers who maintain natural habitats. At the same time, growing awareness of nature conservation issues has resulted in several wetland restoration projects.

Tipperne

At the southern end of Ringkøbing Fjord, a large brackish lagoon of 115 square miles (300 square kilometers) on the west coast of Jutland, lies Tipperne, one of Denmark's oldest and most important wetland reserves. Covering 39 square miles (100 square kilometers), it is one of Denmark's 38 Ramsar Sites and has been designated as an EEC special bird protection area. Tipperne was established to protect breeding birds in the wet meadows, and create a refuge for migrating waterfowl.

However, Tipperne was not always such a haven for

▦ Ramsar Sites

◈ Ramsar Sites/ Parks and Reserves

☐ Water Bodies

☐ Tidal/ Coastal Wetland

▨ Peatland

▽ Deltas

▼ *Lake Kanieris Ramsar site, Latvia. One of the six Ramsar sites in Latvia, Lake Kanieris is one of the country's most important coastal wetlands. It supports about 135 breeding bird species and 700 species of vascular plants.*

Key to numbered Ramsar sites

SWEDEN
1 Umeälv delta

FINLAND
2 Valassaaret-Björkögrunden Archipelago
3 Patvinsuo National Park
4 Porvoonjoki Estuary - Stensböle
5 Vanhankaupunginlahti and Laajalahti Bays
6 Söderskär and Långören Archipelago
7 Aspskär Islands
8 Signilskär-Märket Archipelago
9 Björkör and Lågskär Archipelago

ESTONIA
10 Vilsandi National Park
11 Laidevahe Nature Reserve

12 Hiiumaa Islets and Käina Bay
13 Matsalu Nature Reserve
14 Puhto-Laelatu-Nehatu Wetland Complex
15 Nigula Nature Reserve
16 Soomaa National Park
17 Endla Nature Reserve
18 Alam-Pedja Nature Reserve
19 Muraka Nature Reserve
20 Emajõe Suursoo Mire and Piirissaar Island

LATVIA
21 Northern Bogs (Ziemelu purvi)
22 Lake Engure
23 Lake Kanieris
24 Pape Wetland Complex
25 Lubana wetland complex
26 Teicu and Pelecares bogs

LITHUANIA
27 Kamanos
28 Nemunas Delta
29 Viesvilé

BELARUS
30 Yelnia
31 Osveiski

RUSSIAN FEDERATION
32 Islands in Onega Bay
33 Berezovye Islands, Gulf of Finland
34 Southern coast of the Gulf of Finland, Baltic Sea
35 Svir Delta
36 Kurgalski Peninsula
37 Mshinskaya wetland system
38 Pskovsko-Chudskaya Lowland
39 Oka and Pra River Floodplains

Iceland

The wetlands of Iceland support great numbers of breeding and migratory waterfowl, providing an important staging post in the flyways between the Canadian Arctic and Greenland and western Europe. The wide variety of important wetland habitats includes coastal mud flats, which are particularly important for geese, and inland marshes and lakes, which are essential for breeding ducks and waders. Over 1,800 sq miles (4,600 sq km) are considered to be of international importance and meriting special conservation status.

birds. The region's brackish, wet meadows are a semi-cultivated landscape, created and maintained by farming practices such as hay-making and grazing. Such practices, common up until the 1950s, were productive without fertilizers or draining programmes and they favoured the birds, with hay-making after the birds' breeding period and grazing by cattle and horses during autumn. At that time Tipperne was famous for its large breeding populations of birds such as ruffs (*Philomachus pugnax*), black-tailed godwits (*Limosa limosa*), avocets (*Recurvirostra avosetta*), redshanks (*Tringa totanus*) and gull-billed terns (*Gelochelidon nilotica*). Around 1950, however, a shift in the farming practices in the neighbourhood of Tipperne put an end to the need for hay and grazing areas, with the result that the vegetation grew high and dense, and the breeding populations of the birds dependent on the open wet meadow habitat decreased considerably. In response to this change in the ecological character of the site, subsidies were provided in order to reintroduce farming to the area. Today as many as 50,000 ducks seek protection within the reserve during the hunting season, while 20,000 ducks, 3,000–5,000 geese, and as many as 20,000 shorebirds use the wetland as a feeding area. In spring and summer, large densities of wading birds such as redshanks, black-tailed godwits and avocets have once again returned to Tipperne's meadows.

Western and Central Europe

Stretching from the British Isles and the Brittany Peninsula to the plains of eastern Europe and the shores of the Black and Caspian seas, the middle latitudes of Europe contain some of the most densely industrialized and intensively farmed landscapes in the world. In the wake of the agricultural and industrial revolutions, and the economic development of the centuries that followed, the region has lost the vast majority of its natural wetlands. In western Europe only small areas remain, on the lowland plains or in the high Alps, where montane wetlands are relatively undisturbed. The major exceptions to this, however, are the many estuaries and coastal flats, the floodplains of some of the major rivers, and the wetland mosaic of the Central European Plain. Although it has undergone substantial alterations in many areas, the plain has retained some of the most productive and biologically important ecosystems on the continent.

Coastal wetlands

Much of the European coast consists of a chain of extensive estuaries and intertidal bays separated by long stretches of rocky shore and sandy beaches. The Wadden Sea, which straddles the borders of Denmark, Germany and the Netherlands, is the largest, but areas such as the Wash, Morecambe Bay and the Solway Firth in Britain, and the Baie de Mont St Michel and the Baie de L'Aiguillon in France, are of major international importance. These wetlands are patchworks of sand and mud flats which support large populations of migratory shorebirds. Less visible, but equally important, is the role they play as nurseries for plaice (*Pleuronectes platessa*), sole (*Solea solea*), herring (*Clupea harengus*) and other species of commercially important fish, and their substantial harvest of mussels, cockles and other shellfish.

▲ *Mussels live in dense colonies, attached to rocks by byssus threads.*

Floodplains

All of Europe's major rivers once had extensive floodplains, but today only the Loire, Vistula, and parts of the Danube and Rhine retain a semblance of their former splendour. Dams and dikes to regulate flow and reduce floods have been the major cause of this wetland loss. However, with the growing awareness of the plight these ecosystems face, and a rising debate as to the economic, social and full environmental costs of their loss, many conservation groups in Europe have campaigned actively in recent years against the construction of further dams and river regulation.

▲ *The cockle filter feeds on plankton at high tide using its two siphons.*

Conservation action

The World Wide Fund for Nature (WWF) has been particularly active and played a leading role in encouraging re-examination and cancellation of dams planned for the Danube and Loire, and in encouraging increased conservation investment. Today much of the concern over Europe's floodplains continues to focus on the Danube and its tributaries where plans for improved navigation could lead to the loss of these ecosystems, much as occurred along the Rhine in the 19th and 20th centuries.

The Central European Plain supports a diversity of wetlands, including peat bogs, fish ponds, freshwater marshes, wet meadows and soda lakes as well as the great river floodplains of the Danube, Vistula, Dniepr, Dniester and Volga. Many of these valuable wetlands are still unprotected, although one of the most important sites, the Danube Delta, is now a Biosphere Reserve and listed under both the Ramsar and World Heritage Conventions. On the northern edge of the plain, the Baltic coast supports extensive areas of shallow water and coastal lagoons, which support large numbers of breeding, molting, wintering and migratory birds.

The Rhine

The Rhine River drains an area of 71,000 square miles (185,000 square kilometers) and flows for 790 miles (1,320 kilometers) between the Alps and the North Sea. With a population of well above 50 million people, the Rhine Basin is one of the most densely populated regions in Europe, and is home to several heavily industrialized urban regions such as the Ruhr in Germany. The coal mines, steel mills and oil refineries of these centers have been at the heart of the European economy for much of the past two centuries, and furnished industries producing machinery, electrochemical equipment, automobiles, construction materials, textiles, foodstuffs and paper. Huge chemical factories were also established in the vicinity of Basel in Switzerland, Ludwigshafen, Frankfurt and Cologne in Germany, and Rotterdam in the Netherlands. In addition, a chain of conventional and nuclear power-plants now lines the river and its major tributaries.

Canalization

From the early 1800s onward, the development of this economic heartland along the Rhine led to the loss and degradation of many of the wetlands and aquatic ecosystems of the basin. Most were lost because of huge

■ Ramsar Sites

investment in flood control and improved river navigation. The Upper Rhine floodplain between Basel and Karlsruhe was reshaped from a unique labyrinth of meandering river channels and wooded islands into a dull, monotonous landscape, with straight shipping canals, dikes and drainage ditches. With these drastic changes, the floodplain lost its natural value entirely and its suitability as a habitat for numerous animal and plant communities. Today, dams, weirs and other constructions have made the Upper Rhine and most tributaries inaccessible to migratory fish species such as salmon, which were abundant a century ago.

In addition to the loss of biological diversity, these extensive modifications to the Rhine also reduced the floodplain's role in floodwater retention, thus increasing the risk of flood. Some 60 percent of the floodplain,

6 square miles (15 square kilometers), was lost behind dikes while the newly channelized river took 30 hours instead of 65 to travel from Basel to Karlsruhe. Recognition of this hydrological function and wider appreciation of the role played by the floodplain in purifying water and providing reservoirs for restocking the fish populations of the river has led to increased investment in conserving the remaining floodplains and restoring degraded areas.

River pollutants

In the wake of the postwar economic recovery in western Europe, the Rhine's degradation and loss were compounded by progressive deterioration in water quality. Most types of pollutants showed the same historical trend: a rapid increase in the 1950s and 1960s, peak pollution between 1965 and 1975, and gradual recovery since then. The recovery process is exemplified by decreasing concentrations of organic carbon, ammonium, heavy metals and other pollutants. As a result of diminished inputs of organic substances and ammonium, the oxygen levels improved gradually – an important prerequisite for the recovery of the river fauna.

▼ *The canalization of the Rhine has been an epic civil engineering task spanning several generations. The three diagrams show how, in a period lasting just over 100 years, this section of the Rhine lost the vast majority of its floodplain behind dikes, dams and other constructions.*

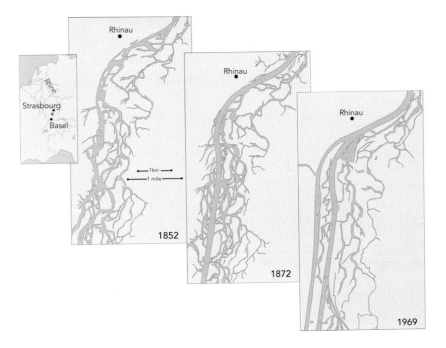

142

■ **Ramsar Sites**

BELARUS
1 Osveiski
2 Yelnia
3 Kotra
4 Olmany Mires Zakaznik
5 Sporovsky Biological Reserve ('zakaznik')
6 Zvanets
7 Mid-Pripyat State Landscape Zakaznik

UKRAINE
8 Shatsk Lakes
9 Prypiat River Floodplains
10 Stokhid River Floodplains
11 Perebrody Peatlands
12 Polissia Mires
13 Desna River Floodplains
14 Lake Synevyr

15 Lower Smotrych River
16 Bakotska Bay
17 Dnipro-Oril Floodplains
18 Kartal Lake
19 Kugurlui Lake
20 Kyliiske Mouth
21 Shagany-Alibei-Burnas Lakes System
22 Shagany-Alibei-Burnas Lakes System
23 Northern Part of the Dniester Liman
24 Dniester-Turunchuk Crossrivers Area
25 Tyligulskyi Liman
26 Yagorlytska Bay
27 Tendrivska Bay
28 Dnipro River Delta
29 Karkinitska and Dzharylgatska Bays
30 Big Chapelsk Depression

31 Central Syvash
32 Molochnyi Liman
33 Obytochna Spit and Obytochna Bay
34 Berda River Mouth and Berdianka Spit and Berdianska Bay
35 Bilosaraiska Bay and Bilosaraiska Spit
36 Kryva Bay and Kryva Spit
37 Eastern Syvash
38 Aquatic-cliff complex of Cape Kazantyp
39 Aquatic-cliff complex of Karadag
40 Aquatic-coastal complex of Cape Opuk

Today, there are many signs that species diversity, namely invertebrates, molluscs and fish, is approaching prewar levels again. This ecological recovery has not been cheap however. In the last three decades, thousands of municipal and industrial sewage treatment plants (STPs) have been built or upgraded. In the German Rhine Basin alone, US$19 billion were spent on sewer and STP construction between 1977 and 1986. In addition, industries have invested large sums in treatment facilities and clean technologies, either voluntarily or under the pressure of new environmental laws.

The Wadden Sea

The Wadden Sea is a shallow coastal wetland stretching for more than 300 miles (500 kilometers), from Den Helder in the Netherlands to the Skallingen Peninsula in Denmark. Covering an area of some 3,000 square miles (8,000 square kilometers), the tidal flats, sandbanks, salt marshes and islands which make up the Wadden Sea constitute Europe's largest intertidal wetland, and are the focus of substantial conservation investment from Denmark, Germany and the Netherlands.

A wildlife haven

Although its size, diversity and transnationality are sufficient cause for international concern, it is the Wadden Sea's role in supporting internationally important populations of animals and plants that have made it one of the world's most famous wetlands. Up to 12 million individuals of over 50 species of waterbird come here in the course of a year, most of them shorebirds from northwest Europe, Scandinavia, Siberia, Iceland, Greenland and northeast Canada. Similarly, most of the 102 species of fish which have been recorded from the Wadden Sea are only found there during certain periods of the year, migrating offshore or along the coast, according to their seasonal and reproductive cycles. Several of these fish are of major commercial importance: 80 percent of plaice, 50 percent of sole, and in some years a large part of the North Sea herring reach maturity in the Wadden Sea.

The common, or harbour seal (*Phoca vitulina*), the grey seal (*Halichoerus grypus*) and the bottle-nose dolphin (*Tursiops truncatus*) are all found in the Wadden Sea. Before 1980, hunting and pollution had exerted considerable pressure upon the population of the harbour seal. However, numbers had increased to 10,000 when, in 1988,

a viral epidemic reduced the population by 60 percent. The population grew again to 21,000 in 2002 but another epidemic of a similar virus reduced the population to 11,000 in August 2003.

Development pressure

The Wadden Sea lies on the coast of one of the most industrialized areas of Europe, and is under considerable environmental pressure. Construction of port facilities and embankments have resulted in damage to, and loss of, important habitats, in particular salt marshes. Although only a modest remainder of the extensive salt and brackish marshes, peatlands and lakes, which covered the area some 2,000 years ago, the salt marshes of the Wadden Sea are still the largest contiguous area of salt marsh in Europe. However, in the 50 years up to 1987, 33 percent of their area was lost to embankments.

Pollution from the rivers which nourish the area, from the North Sea, and from the atmosphere, are of major concern. Of the five rivers entering the Wadden Sea, only the Varde Ao in Denmark is considered a fully natural estuary. The Ems, Weser, Elbe and Elder are all heavily influenced by large-scale engineering, harbour activities and dredging. All four are major sources of both nutrients and contaminants, with the Elbe, Weser and Ems together with the Ijsselmeer discharging some 60 cubic kilometers of polluted water into the Wadden Sea each year. This carries heavy metals, PCBs and pesticides, and large volumes of nutrients.

Resource use and conservation

Mussels have been harvested from the Wadden Sea for

NORTH SEA

Ostfriesische
Inseln
3

1

2

4
5
6

7

Elbe

8   •Hamburg

9

11

10

13

12

Weser   Aller

18

19

BERLIN

14   15   17

16

20

21

Rhein

Köln

22

Elbe

Mulde

23

GERMANY

Frankfurt

Rhein-Main

CZECH
REP.

24

Neckar

Jagst

Donau

Isar

26

25

Donau   Iller

29

28   •München

Inn

27

31

32

30   Bodensee

33   •ZURICH

35

AUSTRIA

34   36   37

38   SWITZERLAND

39   40   41

Léman

Rhône

43

ITALY

10 0   90 km
10 0   60 miles

147

hundreds of years. But today this activity is mainly carried out on culture lots. Fishermen currently harvest seed mussels from other parts of the Wadden Sea and then spread these on the culture lots. This commercial use of the Wadden Sea needs to be managed carefully if it is not to lead to the long term loss of natural mussel banks.

The tourist industry is another important source of income and employment in the Wadden Sea. While this brings many benefits it also threatens to disturb the species that depend on the Wadden Sea. Disturbance of seals during the breeding season is a particular concern as this can reduce nursing time for the calves, leading to increased mortality and reduced breeding success.

In response to these threats to the Wadden Sea concerted international efforts have been mounted to improve the region's conservation status. Since 1978 the governments of The Netherlands, Denmark and Germany, together with a wide range of NGOs, have worked together on the protection and conservation of the Wadden Sea. As a result of these efforts there are signs of progress. The concentrations of pollutants have declined, protected areas have been established in all three countries, and most of the Wadden Sea has been

▲ *Harbor seal* (Phoca vitulina). *One of the most appealing species found in the Wadden Sea, the harbor seal is also one of the most vulnerable. Viral outbreaks have reduced its population dramatically on two occasions in the past 30 years.*

listed under the Ramsar Convention. However there is concern that in some areas the situation is continuing to decline, notably in the estuaries of the Elbe and Weser as a result of deepening of the river channel, and of the Ems as a result of barrage construction.

La Brenne

The Brenne is a land of lakes. Lying in central France, the region boats over 1,100 lakes, covering an area of some 540 square miles (1,400 square kilometers). The lakes are fringed by extensive reedbeds and form one of France's most important, yet little known, wetland systems. Each spring, a wide range or rare and endangered waterbirds, notably bittern (*Botaurus stellaris*), purple heron (*Ardea purpurea*), black-necked grebe (*Podiceps nigricollis*), black tern (*Chlidonias niger*) and bearded tit (*Panurus biarmicus*) nest in the Brenne; while 15 species of amphibian, as well as 60 of the 90 species of dragonfly found in France, also breed here.

In view of the large area of wetland that has been lost because of human activity in other parts of Europe, it is ironic that the lakes of the Brenne are artificial, created by the construction of small dikes to hold back surface water. The first lakes date to the 1100s, when they were built by monks for fish production, most likely carp and pike.

Le Mont St Michel

The bay of Le Mont Saint Michel includes 15 square miles (40 square kilometers) of salt marsh, the largest area on the French coast. Since the 11th century, farmers have used the salt marsh to graze sheep, which have subsequently become famous for the quality of their meat. These "moutons de pré salé;" continue to form an important economic resource today. They graze on

▶ *Black-necked grebe* (Podiceps nigricollis). *Winter plumage (top); breeding plumage (bottom).*

about 9 square miles (24 square kilometers) of the marsh, which they share with migratory birds which visit the bay in winter, and breeding populations in summer. Brent geese (*Branta bernicla*) and widgeon (*Anas penelope*) are the two main species of waterfowl that graze the salt marsh during the winter.

Lac de Grand Lieu

In contrast to the international fame of the Camargue, France's second largest wetland the Lac de Grand Lieu in western France is little known both at home and abroad. Lying to the south of Nantes, the lake stretches over 29 square miles (75 square kilometers) and is bordered by flooded forest of willows and extensive reedbeds. Offshore, an expanse of water lilies alternates with forested islands; true floating islands of trees, each one moving on its muddy base. In spring, a chorus of frogs rings out across the lake, while grey herons (*Ardea cinerea*) and little egrets (*Egretta garzetta*) move between the floating islands where they nest and the surrounding marshes where they feed.

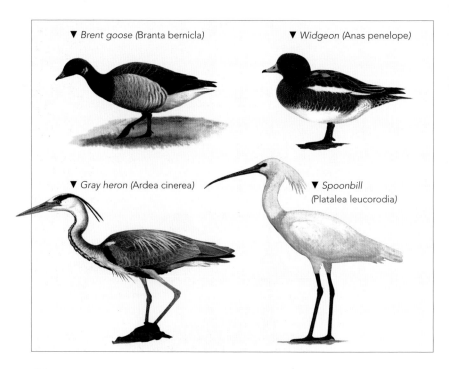

▼ *Brent goose* (Branta bernicla)

▼ *Widgeon* (Anas penelope)

▼ Gray heron (Ardea cinerea)

▼ *Spoonbill* (Platalea leucorodia)

English Channel

Channel Is.

Baie de Somme

Marais du Cotentin et du Bessin, Baie des Veys

Picardie

Aisne

Etangs de la Petite Woëvre

Normandie

Seine

Oise

PARIS

Etangs du Lindre, forêt du Romersberg et zones voisines

Baie du Mont Saint-Michel

Bretagne

Champagne

Etangs de la Champagne humide

Golfe du Morbihan

Basses Vallées Angevines

Grande Brière

Loire

Loire

FRANCE

Bourgogne

Bassin du Drugeon

Marais salants de Guérande et du Més

Lac de Grand-Lieu

La Brenne

Rives du Lac Léman

Marais du Fier d'Ars

Lyon

Lac du Bourget - Marais de Chautagne

ATLANTIC OCEAN

Dordogne

Auvergne

Garonne

Guyenne

Rhône

Gascogne

Canal du Midi

Languedoc

Marseille

Camargue

50   0   175 km

50   0   125 miles

La Petite Camargue

CORSICA

Etang de Biguglia

■ Ramsar Sites

Grand Lieu supports one of Europe's largest colonies of grey herons (1,000 nesting pairs) as well as a colony of European spoonbills (*Platalea leucorodia*). In addition, in winter 20,000 ducks roost on the lake during the day, moving to feed during the night in the Baie de Bourgneuf. A small community of about 20 fishermen use the lake, using nets and traps to capture perch (*Perca fluvialis*), pike (*Esox lucius*), European eel (*Anguilla anguilla*), carp (*Cyprinus carpio*) and tench (*Tinca tinca*), all of which are valued highly locally. In 1980, a nature reserve of 10 square miles (27 square kilometers) was established in the center of the lake, and efforts are under way to expand this further.

The Danube

Draining a basin of 310,800 square miles (805,300 square kilometers), and flowing 2,860 kilometers (1,716 miles) to its delta on the edge of the Black Sea, the Danube is the second largest river in Europe after the Volga. In its lower reaches in Romania, the Danube formerly flooded over 4,250 square miles (11,000 square kilometers) of wetlands, about 5 percent of the country's land surface. These vast wetlands yielded a variety of benefits and played an especially important role in reducing the risk of flood, being able to retain up to 9 cubic kilometers (1.3 cubic miles) of water, and serve as a filter between the basin and the Black Sea.

At 2,000 square miles (5,200 square kilometers), the delta is one of the largest wetland systems in Europe. Characterized by its wilderness of reedbeds, seemingly endless maze of canals, lakes and ponds lined with white willows and poplars, and the fossil dunes with their mosaic of forests and sandy grasslands, the delta forms one of Europe's most important habitats. It supports a rich diversity of plant and animal species, many of which are seriously threatened or have been lost elsewhere in Europe. Bird populations are exceptionally rich, with some 280 species nesting, resting and feeding in the area; 2,500 pairs of pelicans breed each spring together

▼ *Bassin du Drugeon. Lying in the Jura mountains of eastern France the Drugeon valley holds the largest area of bogs in France and a large number of endangered bogland plants. The valley is also the country's most important breeding site for the common snipe (Gallinago gallinago).*

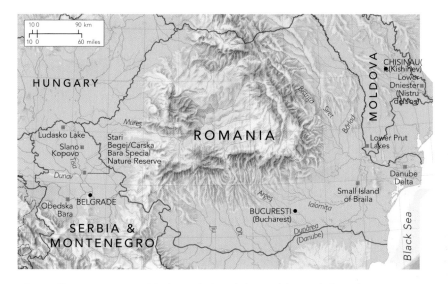

■ Ramsar Sites

with the majority of the world population of pygmy cormorants (*Phalacrocorax pygmeus*)

Modern impacts

Like most large European rivers, the Danube has suffered from pollution, modification of the hydrological system and drainage of its wetlands. Over 1,500 square miles (4,000 square kilometers) – 69 percent – of the floodplains of the lower Danube, and 300 square miles (800 square kilometers) – 15 percent of the delta – have been drained, while dredging of canals within the delta have reduced its capacity to retain and filter water. At the same time industry, agriculture, livestock and urban settlements have all increased the input of sewage, agricultural and industrial waste and pesticides into the Danube's water.

In response to these problems, in 1990 the Romanian Government began to work with international organizations to conserve the biological wealth of the river and its delta, in particular by improving water quality in the basin and restoring some of the former wetland system. The Danube Green Corridor, a regional project to restore floodplains in the Danube basin is also now under way, and in 1991 the delta was listed under both the World Heritage and Ramsar Conventions. However, while these measures have strengthen international support for the delta sustained action at national level continues to be needed. In particular with growing European economic

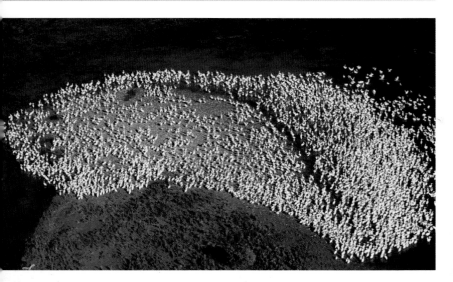

integration pressures on the delta continue. For example in 2004, in the face of international outcry, the Ukrainian Government began to build the Bystroye canal that will improve river transport but threaten the water flow upon which the delta depends.

▲ *Pelican colony in the Danube Delta, Romania. 2,500 pairs of pelicans breed here each year.*

Drainage

In most parts of northwestern Europe, the area of semi-natural grazing marsh and wet meadow has declined sharply in the last 40 years. Much of this has been due to drainage and agricultural improvement. In Belgium, for example, drainage of semi-natural wetlands has taken place at the rate of 10–12 square kilometers (4–5 square miles) a year since 1960. Flooded pastures in river valleys covered about 135 square miles (350 square kilometers) in Flanders in 1960 but have now been reduced to less than 10 percent of this area.

In Germany, as in the Netherlands, most wetlands with potentially productive soils have been drained already, although small wetlands are still under pressure from agricultural improvement. National statistics suggest that the area of bogs and fens within farmland was still falling in the 1980s, being reduced from around 450 square miles (1,170 square kilometers) in 1981 to just over 410 square miles (1,070 square kilometers) in 1985.

In England, landscape surveys show a decline of 52% in the area of freshwater marshes between 1947 and 1982,

■ **Ramsar Sites**
CZECH REPUBLIC
1 Novozámecky a Brehynsky rybník (Novozámeckv/Brehynsky fishponds)
2 Krkonoská raseliniste (Krkonose mountains mires)
3 Libechovka and Psovka Brook
4 Litovelské Pomoraví
5 Poodrí
6 Sumavská raseliniste (Sumava peatlands)
7 Punkva subterranean stream (Podzemní Punkva)
8 Trebonská raseliniste (Trebon mires)
9 Trebonská rybníky (Trebon fishponds)
10 Mokrady dolniho Podyjí (floodplain of lower Dyje River)
11 Lednické rybniky (Lednice fishponds)

SLOVAKIA
12 Wetlands of Orava Basin
13 Turiec Wetlands
14 Rudava River Valley

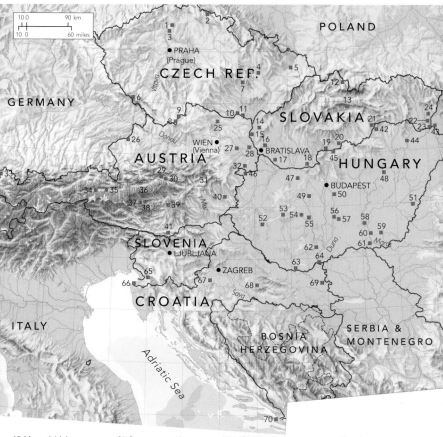

with smaller losses in Wales. Drainage resulted in the annual loss of around 14–28 square miles (40–80 square kilometers) of damp grassland and marsh in England and Wales in the 1970s and early 1980s. Arterial drainage projects, which lower water tables over a substantial area, precipitated some of the most significant habitat losses.

Today the rate of drainage has declined in Belgium, as well as in the United Kingdom, Ireland, the Netherlands and Germany. During the late 1980s and early 1990s there was a sharp fall in the rate of arterial and field drainage, partly because of a cut in the grants available for such work. And where schemes did go ahead, conservation measures took a higher priority than in the past. Today the main concern over wetland drainage is focused on the countries of central and eastern Europe that have recently joined, or will soon be joining, the European Union. Here rapid agricultural development is a major threat to wetlands and there is great concern that these European wetlands with face the same fate as those in much of western Europe in the 20th century.

▼ *The flooded Dyje forest, Czech Republic. Floodplain forests such as these are now among Europe's most endangered wetland habitats. Intensive conservation efforts are now being made to protect these ecosystems and the biodiversity and ecological services that they sustain.*

The Norfolk Broads

The Norfolk Broads are a collection of small lakes along the valleys of the lowland rivers which drain eastern Norfolk. Meaning broadening or widening of a river, the Broads are famous for their reed-fringed rivers and lakes, the gentle contours of the surrounding farmland and the many picturesque thatched cottages. Yet, only a few of the thousands of people who come to visit each year realize that this is a man-made ecosystem. The lakes all lie on peat deposits that were excavated for peat fuel during the 1300s and 1400s. The workings were abandoned when they were flooded due to a rise in sea level and a wetter climate. However, the surrounding marshes and fens continued to be used for hunting of wildfowl, fishing and harvesting reeds (*Phragmites australis*) and sedge (*Cladium mariscus*) for roofing material. This led to the progressive digging out of channels or dikes connecting the lake basins with the rivers so that the marsh products could be transported easily by boat to the surrounding villages and towns.

An invaluable asset

Today, the broads represent many different things to many people. Local people see the wetlands as adding to their household economy; for some they provide fertile land on which to earn a living, while for others they are a source of water and a means of sewage disposal; in addition the region is a place of recreation for those who sail, fish, walk and holiday in its waterways and marshes, while for naturalists it harbours ecosystems which provide a varied plant and animal life.

Over the 19th and 20th centuries, the broads have undergone many environmental changes, most of which have been detrimental to the area. Progressive increase in nutrients in the waters of the broads, combined with widespread loss and degradation of wetland habitats, has led to major changes in plant life of the rivers and broads. The impoverishment of fenland plant communities as a result of neglect and alteration of water cycles has worsened the situation. Further damage has been caused by both the conversion of many of the traditional high-water table grazing marshes into deep-drained arable agriculture, and progressive erosion of the river banks and protective flood walls as a result of boat wash. Uncontrolled mooring and trampling by the hundreds of anglers that visit the broads is also a problem.

In the 1970s, in response to these pressures, the

■ **Ramsar Sites**
UNITED KINGDOM
1 Ronas Hill-North Roe & Tingon
2 East Sanday Coast
3 Caithness & Sutherland Peatlands
4 Caithness Lochs
5 Lewis Peatlands
6 Loch an Duin
7 North Uist Machair and Islands
8 South Uist Machair & Lochs
9 Dornoch Firth & Loch Fleet
10 Loch Maree
11 Cromarty Firth
12 Loch Eye
13 Moray & Nairn Coast
14 Loch Spynie
15 Loch of Strathbeg
16 Ythan Estuary & Meikle Loch
17 Loch Ruthven
18 Loch of Skene
19 River Spey-Insh Marshes
20 Cairngorm Lochs
21 Muir of Dinnet
22 Coll
23 Claish Moss
24 Inner Moray Firth
25 Loch of Lintrathen
26 Montrose Basin
27 Sleibhtean agus Cladach Thiriodh (Tiree Wetlandsand Coast)
28 Rannoch Moor
29 Firth of Tay and Eden Estuary
30 Loch of Kinnordy
31 South Tayside Goose Roosts
32 Loch Leven
33 Cameron Reservoir
34 Loch Lomond
35 Firth of Forth
36 Gruinart Flats, Islay
37 Rinns of Islay
38 Bridgend Flats, Islay
39 Eilean na Muice Duibhe (Duich Moss), Islay
40 Inner Clyde Estuary
41 Westwater
42 Gladhouse Reservoir
43 Fala Flow
44 Greenlaw Moor
45 Kintyre Goose Roosts
46 Din Moss-Hoselaw Loch
47 Silver Flowe
48 Castle Loch,

Lochmaben
49 Loch of Inch & Torrs Warren
50 Loch Ken & River Dee Marshes
51 Holburn Lake & Moss
52 Lindisfarne
53 Northumbria Coast
54 Irthinghead Mires
55 Upper Solway Flats & Marshes
56 Esthwaite Water
57 Teesmouth & Cleveland Coast
58 Duddon Estuary
59 Morecambe Bay
60 Leighton Moss
61 Malham Tarn
62 Lower Derwent Valley
63 Ribble & Alt Estuaries
64 Martin Mere
65 Humber Flats, Marshes & Coast (Phase 1)
66 Mersey Estuary
67 Rostherne Mere
68 Gibralter Point
69 Midland Meres & Mosses (Phase 2)
70 Midland Meres & Mosses (Phase 1)
71 Rutland Water
72 The Wash
73 North Norfolk Coast
74 Dersingham Bog
75 Roydon Common
76 Broadland
77 Woodwalton Fen
78 Ouse Washes
79 Breydon Water
80 Nene Washes
81 Wicken Fen
82 Chippenham Fen
83 Redgrave & South Lopham Fens
84 Minsmere-Walberswick
85 Alde-Ore Estuary
86 Deben Estuary
87 Stour & Orwell Estuaries
88 Abberton Reservoir
89 Colne Estuary (Mid-Essex Coast Phase 2)
90 Hamford Water
91 Walmore Common
92 Blackwater Estuary
93 Crouch & Roach Estuaries (Mid-Essex Coast Phase 3)
94 Dengie (Mid-Essex Coast Phase 1)
95 Lee Valley
96 Foulness (Mid-Essex

Coast Phase 5)
97 Severn Estuary
98 Thames Estuary and Marshes
99 Benfleet & Southend Marshes
100 South West London Waterbodies
101 Medway Estuary & Marshes
102 The Swale
103 Thanet Coast & Sandwich Bay
104 Thursley & Ockley Bogs
105 Stodmarsh
106 Somerset Levels & Moors
107 Isles of Scilly
108 Exe Estuary
109 Chesil Beach & The Fleet
110 Dorset Heathlands
111 Poole Harbour
112 Avon Valley
113 New Forest
114 Solent and Southampton Water
115 Portsmouth Harbour
116 Chichester & Langstone Harbours
117 Pagham Harbour
118 Arun Valley
119 Pevensey Levels
120 South East Coast of Jersey, Channel Islands
121 Corsydd Môn a Llyn (Anglesey & Llyn Fens)
122 Llyn Idwal
123 The Dee Estuary
124 Llyn Tegid
125 Cors Fochno & Dyfi
126 Cors Caron
127 Burry Inlet
128 Crymlyn Bog
129 Garry Bog
130 Lough Foyle
131 Garron Plateau
132 Black Bog
133 Larne Lough
134 Ballynahone Bog
135 Fairy Water Bogs
136 Lough Neagh & Lough Beg
137 Belfast Lough
138 Strangford Loch
139 Pettigoe Plateau
140 Fardrum and Roosky Turloughs
141 Slieve Beagh
142 Cuilcagh Mountain
143 Upper Lough Erne
144 Turmennan Lough
145 Carlingford Lough

IRELAND
146 Trawbreaga Bay
147 Lough Barra Bog
148 Meenachullion Bog
149 Pettigo Plateau
150 Blacksod Bay and Broadhaven
151 Knockmoyle/Sheskin
152 Killala Bay/Moy Estuary
153 Cummeen Strand
154 Owenduff catchment
155 Owenboy
156 Easky Bog
157 Lough Gara
158 Lough Oughter
159 Dundalk Bay
160 Lough Glen
161 Lough Corrib
162 Lough Derravaragh
163 Lough Iron
164 Lough Owel
165 Broadmeadow Estuary
166 Lough Ennell
167 Rogerstown Estuary
168 Baldoyle Bay
169 Mongan Bog
170 Sandymount Strand/Tolka Estuary
171 Inner Galway Bay
172 Coole Lough & Garryland Wood
173 Clara Bog
174 Raheenmore Bog
175 Ballyallia Lough
176 Slieve Bloom Mountains
177 Pollardstown Fen
178 North Bull Island
179 Tralee Bay
180 Castlemaine Harbour
181 Gearagh, The
182 Cork Harbour
183 Ballycotton Bay
184 Ballymacoda
185 Blackwater Estuary
186 Dungarvan Harbour
187 Tramore Backstrand
188 Raven, The
189 Bannow Bay
190 Wexford Wildfowl Reserve

Broads Authority embarked on a restoration program and this continues today. The major focus of much of this work has been on reducing nutrient flow into the Broads, with special emphasis placed on stripping phosphate from sewage before it enters the Broads. As a result of these efforts substantial progress is being made. A study by English Nature and the Broads Authority has shown that by 2004 9% of the open water had been restored to ecological health and a further 50% is expected to have reached this condition by 2010. Upton, Buckenham and the two Martham Broads are some of the areas that have been restored to health.

▲ *Reed harvest in the Norfolk Broads. Until the 20th century many of Britain's reedbeds were intensively managed for harvest of thatching reed. While use of tiles reduced the demand for reed and led to the conversion of reedbeds to other uses, there is still a significant demand for reeds today.*

The Mediterranean Basin

When the empires of ancient Greece and Rome ruled much of the Mediterranean it was a region rich in wetlands. The wetlands remained largely intact until the 1800s and 1900s, when most were drained either to provide agricultural land or to eradicate malaria. Today, the remaining wetlands of the Mediterranean comprise an estimated 2,300–3,300 square miles (6,000–8,500 square kilometers) of coastal lagoons, between 3,000 and 3,850 square miles (8,000–10,000 square kilometers) of natural lakes and marshes, mainly lying in river deltas and the region's remaining floodplains, and over 3,850 square miles (10,000 square kilometers) of artificial wetlands, mostly reservoirs.

The deltas of the north shore of the Mediterranean support some of the most extensive and varied wetland areas remaining in the whole of the Mediterranean Basin. The Ebro in Spain, Rhône in France, Po in Italy, Axios and Evros in Greece, Menderes, Seyhan and Ceyhan in Turkey are all complex mosaics of wetland and dryland habitats, including dunes with juniper and pine trees, lagoons, freshwater and brackish marshes, and freshwater lakes.

Along the southern Mediterranean shore of North Africa, most rivers are short, highly seasonal and do not provide sufficient sediment for delta formation, the Nile being the only exception. However, the flow of sediment to the delta has been reduced greatly by the Aswan High Dam (completed in 1971) and its predecessor (completed in 1902). Inland, other wetlands, such as sebkhas and oases exist. Sebkha El Kelbia in Tunisia, Tamentit Oasis in Algeria, and Siwa Oasis in Egypt, are characteristic examples.

Vanishing wetlands

Coastal lagoons, such as those of the Languedoc in southern France, are among the most typical wetlands of the Basin and are scattered in an irregular fashion along its shores. Most are connected to the sea by narrow channels, which are either artificial or natural, permanent or temporary, and vary in salinity from season to season.

Riverine floodplains have been reduced to a few tiny isolated remnants as the result of 2,000 years of hydraulic engineering. Oxbows and floodplain freshwater marshes are now rare as most have long since been drained for agriculture. The only surviving examples are found in a fragmentary and highly modified state in the Languedoc and Crau regions in southern France, in the Po Valley in northern Italy, in northern Algeria, and in the now much degraded Parque Nacional de las Tablas de Daimiel on the River Guadiana in central Spain. Along with the destruction of the floodplains, diking, grazing, agriculture and felling for timber have killed the riverine forest. Only a few isolated relict stands still exist in the Mediterranean region, such as Kotza Orman in the Nestos Delta in northern Greece and around the Moraca River in Yugoslavia.

Natural freshwater lakes are also infrequent in the Mediterranean region except for a few examples such as Lago Maggiore and Lago di Garda in Italy, Lakes Mikri and Megali Prespa shared by Albania, Greece and Macedonia, and others in Yugoslavia, Albania and Turkey. In North Africa, permanent freshwater lakes have

■ **Ramsar Sites**
SPAIN
1 Laguna y Arenal de Valdoviño
2 Rías de Ortigueira y Ladrido
3 Ría del Eo
4 Marismas de Santoña
5 Ria de Mundaka-Guernika
6 Txingudi
7 Colas del Embalse de Ullibarri
8 Salburua
9 Lago de Caicedo-Yuso y Salinas de Añana
10 Lagunas de Laguardia (Alava): Carralogroño, Carravalseca, Prao de la Paul y Musco
11 Embalse de las Cañas
12 Laguna de Pitillas

always been scarce outside of the subhumid mountainous areas, and the largest examples have now been drained, such as Lake Fetzara in Algeria. The only large lowland freshwater lakes that remain either dry out periodically or are in coastal areas and, such as Lake Ichkeul in Tunisia, receive some inflow of sea water during the drier periods of the year and in times of drought. Most are seriously affected by hydraulic works.

Several countries have inland salt lakes. These vary from sites lying below present sea level such as Chott Melrir in Algeria, to upland basins at over 3,300 feet (1,000 meters) such as the Plain of Chotts in Algeria, the Laguna de Gallocanta in Spain and Lake Tuz in Turkey. All of these salt lakes are shallow and dry out completely from time to time.

Conservation issues

Just about every type of wetland in the Mediterranean region has been exploited by man since the beginning of civilization. Among the most common human uses have been drainage and conversion to agriculture, grazing, water storage, fisheries and aquaculture, mineral exploitation, hunting, harvesting of wetland vegetation, tourism and water sports. In recent years, urban and industrial development have also increased substantially, including transportation infrastructure (airports, harbors and roads), adding greatly to the rate of wetland loss, much of it through drainage for building land.

With rapid economic development in many countries and increasing population pressures in the south and east, compounded by mass tourism, the present demand for water now exceeds available resources in

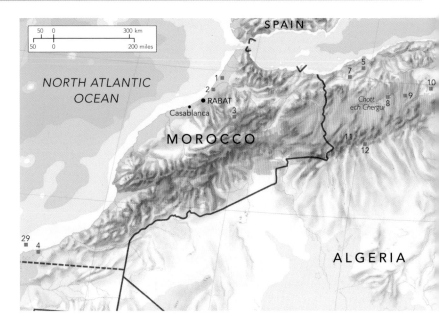

many parts of the Mediterranean. As a result the future management of freshwater within the basin and the degree to which this takes account of conservation requirements is now the single greatest challenge facing the wetlands of the region.

However the countries of the basin have made an explicit commitment to wetland conservation by ratifying the Ramsar Convention. The last two to sign the Convention were Libya (2000) and Bosnia and Herzegovina (2001). Still, the future for Mediterranean wetlands and their flora and fauna is uncertain. Better understanding of the services and values provided by wetlands and far greater investment in conservation will be needed if the increasing public and government interest is to halt the degradation and loss of these invaluable habitats.

Flamingos of the Camargue

The Rhône Delta region, better known as the Camargue, is host to hundreds of thousands of visitors each year; some are on vacation, others on day trips to the beach. As they drive across this flat landscape most hope to see the black bulls, white horses and, above all, the "pink" or greater flamingos (*Phoenicopterus ruber roseus*), which, in the mind of most visitors, characterize this most

■ **Ramsar Sites**
MOROCCO
1 Merja Zerga
2 Merja Sidi Boughaba
3 Lac d'Afennourir
4 Baie de Khnifiss

ALGERIA
5 Marais de la Macta
6 Réserve Naturelle du
  Lac de Réghaïa
7 Sebkha d'Oran
8 Chott Ech Chergui
9 Grotte karstique de
  Ghar Boumâaza
10 Chott de Zehrez
   Gharbi
11 Oasis de Moghrar et
   de Tiout
12 Le Cirque de Aïn
   Ouarka
13 Chott de Zehrez
   Chergui
14 Chott El Hodna
15 Réserve Naturelle du
   Lac de Béni-Bélaïd
16 Complexe de zones
   humides de la plaine
   de Guerbes-
   Sanhadja
17 Tourbière du Lac
   Noir

18 Réserve Intégrale du
   Lac Tonga
19 La Réserve Naturelle
   du Lac des Oiseaux
20 Réserve Intégrale du
   Lac Oubeïra
21 Lac de Fetzara
22 Marais de la
   Mekhada
23 Aulnaie de Aïn Khiar
24 Chott Melghir
25 Chott Merrouane et
   Oued Khrouf
26 Oasis de Ouled Saïd
27 Oasis de Tamantit et
   Sid Ahmed Timmi

TUNISIA
28 Ichkeul

SPAIN
29 Saladar de Jandía

▲ *Flamingos of the Camargue. Salt pans provide a stable supply of algae and invertebrates on which flamingos feed with their specially adapted straining beaks, while* *also providing a secure artificial nesting site. These near perfect conditions have led to a sharp increase in the numbers of breeding birds in recent decades.*

famous of French wetlands. Flamingos have become a symbol of conservation success in France. Since records began in 1944, the number of flamingos breeding in the Camargue has grown from about 3,000 pairs in 1947 to more than 20,200 in 2000. Since 1981, at least 8,000 pairs have nested each year. In winter too, the population has risen, with the number wintering in France growing from 500 in 1969 to 20–30,000 each winter since 1991.

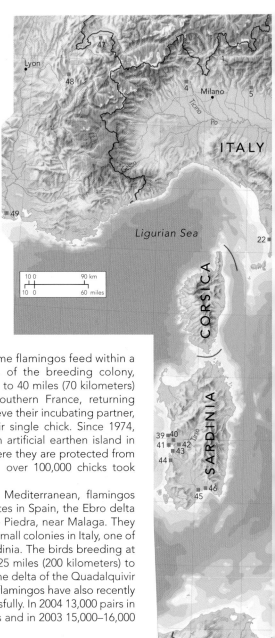

The reasons for this increase in both breeding and wintering populations are complex. However, a central factor has been the presence of a secure nesting and feeding site in the salinas (salt pans and basins) of the Camargue. Although some flamingos feed within a 6-mile (10-kilometer) radius of the breeding colony, others fly to feeding sites up to 40 miles (70 kilometers) away along the coast of southern France, returning every two to four days to relieve their incubating partner, and in due course feed their single chick. Since 1974, breeding birds have used an artificial earthen island in the Etang de Fangassier, where they are protected from disturbance and from which over 100,000 chicks took wing between 1974–2000.

Elsewhere in the western Mediterranean, flamingos have bred regularly at two sites in Spain, the Ebro delta and the lagoon of Fuente de Piedra, near Malaga. They also breed regularly in three small colonies in Italy, one of which is on the island of Sardinia. The birds breeding at Fuente de Piedra fly up to 125 miles (200 kilometers) to feed in the Coto Doñana in the delta of the Quadalquivir River, south of Seville, where flamingos have also recently begun to breed again successfully. In 2004 13,000 pairs in 7 colonies raised 3,500 chicks and in 2003 15,000–16,000 pairs raised 1,200 chicks.

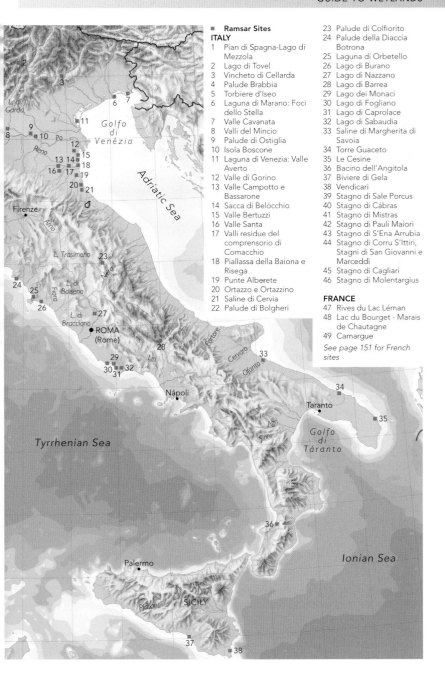

■ **Ramsar Sites**

ITALY
1 Pian di Spagna-Lago di Mezzola
2 Lago di Tovel
3 Vincheto di Cellarda
4 Palude Brabbia
5 Torbiere d'Iseo
6 Laguna di Marano: Foci dello Stella
7 Valle Cavanata
8 Valli del Mincio
9 Palude di Ostiglia
10 Isola Boscone
11 Laguna di Venezia: Valle Averto
12 Valle di Gorino
13 Valle Campotto e Bassarone
14 Sacca di Belócchio
15 Valle Bertuzzi
16 Valle Santa
17 Valli residue del comprensorio di Comacchio
18 Piallassa della Baiona e Risega
19 Punte Alberete
20 Ortazzo e Ortazzino
21 Saline di Cervia
22 Palude di Bolgheri

23 Palude di Colfiorito
24 Palude della Diaccia Botrona
25 Laguna di Orbetello
26 Lago di Burano
27 Lago di Nazzano
28 Lago di Barrea
29 Lago dei Monaci
30 Lago di Fogliano
31 Lago di Caprolace
32 Lago di Sabaudia
33 Saline di Margherita di Savoia
34 Torre Guaceto
35 Le Cesine
36 Bacino dell'Angitola
37 Biviere di Gela
38 Vendicari
39 Stagno di Sale Porcus
40 Stagno di Cábras
41 Stagno di Mistras
42 Stagno dei Pauli Maiori
43 Stagno di S'Ena Arrubia
44 Stagno di Corru S'Ittiri, Stagni di San Giovanni e Marceddì
45 Stagno di Cagliari
46 Stagno di Molentargius

FRANCE
47 Rives du Lac Léman
48 Lac du Bourget - Marais de Chautagne
49 Camargue

See page 151 for French sites

167

The salinas

Salt has for centuries been an important trading commodity and the basis of major trade routes. One of the easiest ways to obtain salt is to allow sea water to evaporate naturally using the heat of the sun and wind, and then collect the residue. This technique is only really economically viable in areas with long periods of warm weather and little precipitation – the long Mediterranean summer is ideal and – since Antiquity – many salines produce salt for the table and for industry in this way.

Salinas are composed of basins and salt pans for the production of salt; in total, they cover some 400 square miles (1,000 square kilometers) and exist in virtually all countries of the Mediterranean region. Some are run as state or privately owned industries, others are smaller, traditional and less efficient. The largest salina covers 42 square miles (110 square kilometers) and lies in the south of the Camargue. In most salinas sea water is moved, or pumped, through a series of pans, becoming increasingly saltier due to evaporation as it approaches the final settling pans where the salt crystallizes. This circuit might take anything up to two years to complete in a large salina.

As the water moves through the system, the regular flooding and drying of the shallow pan creates ideal conditions for certain algae and invertebrates, providing rich feeding grounds for birds. The brine shrimp *Artemia*, for example, is particularly important, and is a key component of the diet of many bird species, in particular flamingos. In addition to their high productivity, the presence of many ponds of different salinity at any one time allows many different species of bird to coexist,

■ Ramsar Sites

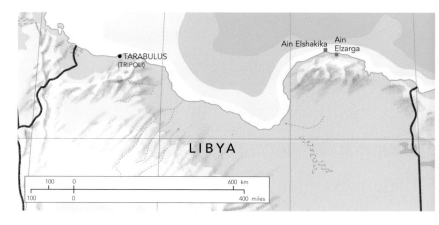

while the stability of the pumping cycle from year to year means that conditions are usually predictable. This stability and predictability are of special value in the Mediterranean, where wetland conditions tend to be extremely variable and the lack of a strong tidal cycle means that there are few mud flats. For these reasons salinas are key breeding and feedings sites for flamingos, avocets (*Recurvirostra avosetta*), little terns (*Sterna albifrons*), common terns (*Sterna hirundo*), audouin's gulls (*Larus uadouinii*), slender-billed gulls (*Larus genei*), Kentish plovers (*Charadrius alexandrinus*), gull-billed terns (*Gelochelidon nilotica*) and sandwich terns (*Sterna sandvicensis*). Between 50 and 80 percent of all flamingos in southern France feed in salinas during the summer months.

Salinas are a good example of how economic interests can coexist with, and even benefit, local wildlife. However, small salinas and even some larger enterprises are increasingly uneconomic and their closure may have considerable local consequences for bird populations. In Portugal, for example, two-thirds of the 300 salinas are either already closed, or threatened with closure, while in France the commercial future of the salinas in the Camargue is uncertain. Conservationists do not at present have the resources to keep salinas functioning purely for their ornithological interest and importance, and this is a source of concern for the future of the bird populations that use them and the spectacle they provide. However, a number of small, traditional salinas continue to be viable by producing a highly quality product; Sekoveljske saline in Slovenia is a good example of this.

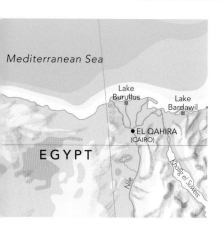

Traditional fisheries and aquaculture

Rivers, reservoirs, lakes and coastal lagoons contribute about 25 percent of the total commercial fishing undertaken within the Mediterranean Basin. Of these four wetland types, lakes and coastal lagoons are the most important, with the largest number of fish caught in lagoons. Fishing in lagoons is generally seasonal, taking place in autumn and spring. The main target species include eels, grey mullet (*Mullus* sp.), sea bass (*Dicentrarchus labrax*) and sea bream (*Pagellus bogaraveo*), all of

169

which spawn in the sea and use the relative safety of the lagoons for feeding and growth.

■ Ramsar Sites

Over the past 20 years, these traditional fisheries have declined and aquaculture has become increasingly important. The main shellfish species farmed are mussels, oysters and clams, although, more recently, cage culture of sea bass and sea bream, and pond culture of prawns and shrimps has been developed. This has changed the economy and ecological character of many of the region's coastal lagoons, and there are concerns about the long-term environmental impact and sustainability of this use.

Turkey's wetland diversity

Surrounded by water on three sides – the Black Sea to the north and the Mediterranean to the south and west – Turkey is home to some of the most important wetland resources in the Mediterranean. Two large deltas are situated on the Mediterranean coast, one on the Aegean, and another on the Black Sea. Most coastal wetlands consist of extensive lagoon systems that are not only important for wildfowl but also for the valuable local fisheries. Inland, Turkey has a variety of wetland types. There are, for example, many deep freshwater lakes – especially in the "lake district" of western Anatolia; the central plateau has extensive marshes and shallow lakes, while eastern Turkey is a region of soda lakes and river meadows. One of the country's most important lakes is Burdur Golu, which in winter supports up to 70 percent of the world's population of white-headed duck (*Oxyura leucocephala*).

On the south coast of Turkey, the Goksu Delta is one of the country's most important remaining wetlands. Covering some 37,000 acres it is home to the majority of the Turkish population of purple gallinule (*Porphyrio porphyrio*) and a large breeding population of the marbled teal (*Marmarcetta angustirostris*). The delta is also an important stop-over for white storks (*Ciconia ciconia*) and white pelicans (*Pelecanus onocrotalus*), white-tailed eagles (*Haliaeetus albicilla*) and spotted eagles (*Aquila clanga*), as well as tens of thousands of wintering waterfowl. The delta became a protected area in 1990 and conservation efforts have intensified over the past 15 years.

Greece

Greek mythology, archaeological research, and historical records all show that the economic, political, social,

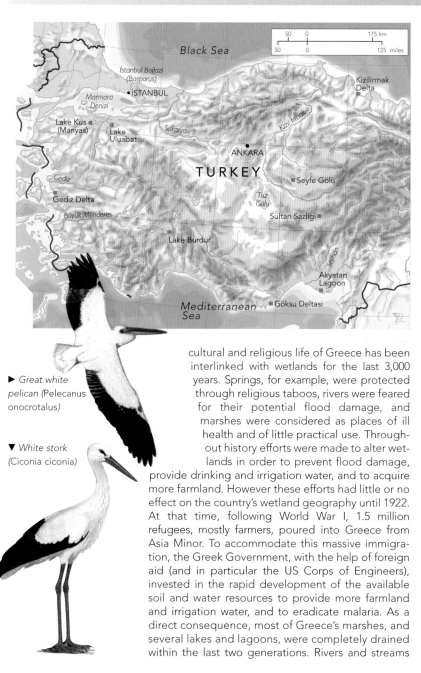

Black Sea

İstanbul Boğazı
(Bosporus)

Marmara
Denizi

•İSTANBUL

Kizilirmak
Delta

Lake Kus
(Manyas)

Lake
Uluabat

Sakarya

Kizil Irmak

•ANKARA

TURKEY

Gediz

•Seyfe Gölü

Gediz Delta

Tuz
Gölü

Büyük Menderes

Sultan Sazligi

Lake Burdur

Seyhan

Akyatan
Lagoon

Mediterranean
Sea

•Göksu Deltasi

50   0   175 km
50   0   125 miles

▶ *Great white
pelican (Pelecanus
onocrotalus)*

▼ *White stork
(Ciconia ciconia)*

cultural and religious life of Greece has been interlinked with wetlands for the last 3,000 years. Springs, for example, were protected through religious taboos, rivers were feared for their potential flood damage, and marshes were considered as places of ill health and of little practical use. Throughout history efforts were made to alter wetlands in order to prevent flood damage, provide drinking and irrigation water, and to acquire more farmland. However these efforts had little or no effect on the country's wetland geography until 1922. At that time, following World War I, 1.5 million refugees, mostly farmers, poured into Greece from Asia Minor. To accommodate this massive immigration, the Greek Government, with the help of foreign aid (and in particular the US Corps of Engineers), invested in the rapid development of the available soil and water resources to provide more farmland and irrigation water, and to eradicate malaria. As a direct consequence, most of Greece's marshes, and several lakes and lagoons, were completely drained within the last two generations. Rivers and streams

171

■ **Ramsar Sites**
SERBIA AND
MONTENEGRO
1 Skadarsko Jezero
For other sites see map on
page 153

BULGARIA
2 Ibisha Island
3 Belene Islands Complex
4 Srebarna
5 Durankulak Lake
6 Lake Shabla
7 Atanasovo Lake
8 Pomorie Wetland
   Complex
9 Vaya Lake
10 Poda
11 Ropotamo Complex

**MACEDONIA (THE
FORMER YUGOSLAV
REPUBLIC OF)**
12 Lake Prespa

ALBANIA
13 Karavasta Lagoon
14 Butrint

have been dammed, diked and re-routed, and floodplain forests felled. In total, some two-thirds of the wetland area of the country has been lost, with a consequent decline in the country's biological diversity.

Conservation support

In spite of this destruction, Greece still possesses some of the Mediterranean's most significant wetlands, especially in Macedonia and Thrace, and along the western coast. In total, more than 400 wetland sites – small and large – have been identified, all of which merit conservation action in the face of continuing agricultural, industrial and tourist development.

In response to the growing concern for wetlands in Greece, WWF and the European Commission helped the Ministry of Environment and the Goulandris Museum to

172

▲ *Prespa Lake lies on the border between Albania, Greece, and the Republic of Macedonia. The lake is an important breeding site for waterbirds, including 500 pairs of the Dalmatian pelican* (Pelecanus crispus) *and over 100 pairs of the great white pelican (P.* onocrotalus).

establish in 1991 the Greek Biotope/Wetland Center (EKBY). The center started work in 1991 and is located in Thessaloniki, the capital of Macedonia, and close to some of Greece's most important wetlands, such as Lakes Volvi and Coronia, the Axios River Delta and Lake Kerkini. The focus of the center is to improve understanding of the country's wetlands, monitor their status, raise awareness within government and the public of the pressures facing wetlands, and encourage action to redress the situation. In northwestern Greece, the Society for the Protection of Prespa has played a major role in convincing local people of the value of their wetlands (the two Prespa lakes) and has been instrumental in the establishment of the transboundary Prespa Park. Despite these successes 7 of the 10 Greek Ramsar sites have undergone profound – and negative – ecological changes. Sustained efforts are therefore required if Greece's wetlands are to continue to exist.

The Middle East

Lower Mesopotamia, the land between the Tigris and Euphrates rivers, is the legendary site of the Garden of Eden. Here, on the margins of the largest continuous wetland system in the Middle East, civilization was established as early as 4,000 years BC, and a sophisticated irrigation system was developed. Until the early 1990s these marshes still covered 6,000 square miles (15,000 square kilometers) in what is now southern Iraq, their luxuriant growth contrasting with the arid landscape of most of the Middle East.

This complex of shallow freshwater lakes and marshes, river channels and seasonally inundated floodplains, together with most other wetlands of the region, owes its existence to the mountain ranges of eastern Turkey, northern Iran and Afghanistan. These receive sufficient rain and snow to feed several major river systems, which in turn sustain several impressive wetlands, including the lakes of the Rift Valley in the Levant and Jordan, the vast marshes of the Tigris and Euphrates in Mesopotamia, the wetlands of the Orumiyeh Basin, South Caspian lowlands, Khuzestan and central Fars in Iran, and the wetlands of the Seistan Basin on the Iran/Afghanistan border.

▼ *Lumps of crystallized salt litter the surface of the Dead Sea in Israel. The extreme salinity of this inland sea places it among those very few wetlands which are virtually lifeless.*

174

Although the Mesopotamian marshes of southern Iraq are the most extensive wetland ecosystems in the Middle East, a similar but much smaller complex of wetlands lies in neighboring southwestern Iran. Three large rivers, the Karun, Dez and Kharkeh, rise in the Zagros Mountains. These flow out onto the plains of Khuzestan, creating an area of seasonal floodplain wetlands, which extend south to the head of the Gulf. In northern Iran, the marshes of the South Caspian lowlands are equally impressive. Here there is an almost unbroken chain of freshwater lakes and marshes, brackish lagoons, irrigation ponds (locally known as "ab-bandans") and rice paddies stretching for

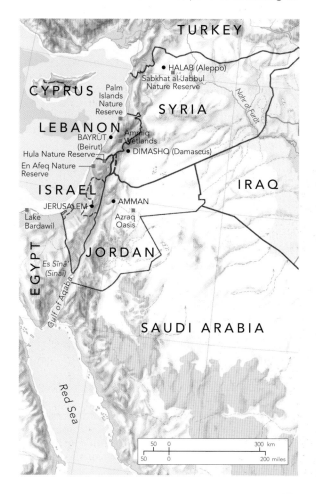

■ Ramsar Sites

over 400 miles (700 kilometers) along the Caspian shore. Fed by abundant rainfall on the humid north slope of the Alborz Mountains, most of these wetlands are permanent and support vegetation all year. The two largest wetlands in this region are the Anzali Mordab complex in the southwest Caspian, and the Gorgan Bay/Miankaleh Peninsula complex in the southeast.

Wetland diversity

Many of the Middle East's major rivers lie in internal drainage systems, ending up in land-locked, brackish to saline lakes which are often subject to wide fluctuations in water level and may dry out completely during periods of drought. Extensive fresh to brackish marshes occur where rivers and spring-fed streams enter several of these salt lakes. Sabkat al Jabboul in Syria, the Dead Sea in the Jordan Valley, Lake Orumiyeh, Gavkhouni Lake and the Neiriz Lakes in Iran, and the wetlands of the Seistan Basin are all examples. The latter are unusual in that although the three main lakes of the Seistan Basin, namely Hamoun-i Puzak, Hamoun-i Sabari and Hamoun-i Helmand, lie within an internal drainage basin, they are predominantly freshwater. The system is fed by the Helmand River, which rises in the Hindu Kush in northern Afghanistan. In most years, the Helmand supplies sufficient water to flood only one or two of the lakes, but about once every decade, the floodwaters of the Helmand sweep through all three lakes and overflow into a vast salt waste to the southeast, flushing the salts out of the system.

Further south in the Middle East, there are very few natural wetlands away from the coast. Azraq Oasis, in Jordan's Eastern Desert, is a notable exception, but this has now been reduced greatly by extraction of groundwater. In the extremely arid interior of the Arabian Peninsula, the only significant natural wetlands are the *playa* wetlands (or "qa") of northern Saudi Arabia – low-lying desert areas which are temporarily flooded during periods of heavy winter rainfall. In some parts of Yemen, subsurface seepage feeds saline grazing marshes in valley bottoms, while in southwestern Oman, the slight influence of the southwest monsoon provides sufficient rainfall to maintain a chain of brackish creeks and lagoons along the narrow coastal plain.

By contrast, the coastline of southern Iran and the Arabian Peninsula possesses extensive wetlands. Intertidal mud flats run along the coasts of the Gulf and

■ **Ramsar Sites**

GEORGIA
1 Wetlands of Central
   Kolkheti
2 Ispani II Marshes

ARMENIA
3 Lake Arpi
4 Lake Sevan

AZERBAIJAN
5 Agh-Ghol
6 Ghizil-Agaj
7 Kirov Bays

TURKMENISTAN
8 Krasnovodsk &
   North-Cheleken Bays

IRAN
9 Lake Urmia [or Orumiyeh]
10 Lake Gori
11 Shurgol, Yadegarlu &
   Dorgeh Sangi Lakes
12 Lake Kobi
13 Bandar Kiashahr Lagoon
   & mouth of Sefid Rud
14 Anzali Mordab (Talab)
   complex
15 Amirkelayeh Lake
16 Fereydoon Kenar,
   Ezbaran & Sorkh Ruds
   Ab-Bandans
17 Miankaleh Peninsula,
   Gorgan Bay & Lapoo-
   Zaghmarz Ab-bandan
18 Gomishan Lagoon
19 Alagol, Ulmagol & Ajigol
   Lakes

southern Red Sea, and also in some areas of the Arabian Sea coast. They are very extensive at the head of the Gulf in Iran, Iraq and Kuwait, along the southern shore of the Gulf in Saudi Arabia and the Gulf States, and at Barr-al-Hekman on the Oman coast. Many of these mud flats support simple mangrove communities consisting of only a single species, *Avicennia marina*. Much the largest stands of mangroves grow in the Khuran Straits in southern Iran, where there are some 400 square miles (1,000 square kilometers) of low-lying islands with broad mangrove fringes and extensive intertidal mud flats.

Competition for water

The wetlands of the Middle East have suffered from a range of human impacts, including drainage, pollution and in-fill for urban and industrial development. However,

177

in a region in which water is a scarce resource almost everywhere, wetlands have come under particularly severe pressure. Throughout Iran, Iraq, Syria, Lebanon, Israel and Jordan, rivers have been diverted for irrigation purposes, and domestic and industrial consumption, while the wetlands themselves have been drained for agriculture, industry and urban development. The increasing utilization of the waters of the Tigris and Euphrates rivers for irrigation in Turkey, Syria and upper Iraq had resulted in a significant loss of wetlands in Mesopotamia even prior to the drainage of the Basra marshes by Saddam Hussein after the first Gulf War. Flood control projects and irrigation schemes on the Helmand River in Afghanistan have also had an adverse effect on the wetlands of the Seistan Basin in Iran, especially during years of below-average rainfall.

In Syria, Lebanon and Israel, almost all of the original freshwater wetlands were drained for agriculture at the beginning of the 1900s, while in the Jordan Valley, the level of the Dead Sea has fallen dramatically as a result of water being diverted from the Jordan River for irrigation schemes.

The Euphrates under threat

The largest water management scheme in the region is an ambitious project to harness the Euphrates to produce power and irrigate over 70 square miles (185 square kilometers) of arid land in Turkey's southeast provinces. However, full development of the scheme could mean that Syria would lose 40 percent of its Euphrates water and Iraq up to 90 percent. A total of 21 schemes have been planned to irrigate some 42,000 square miles (108,000 square kilometers), in southeast Turkey, a land area the size of Cuba.

The Marsh Arabs

Undoubtedly the best-known inhabitants of the wetlands of lower Iraq are the Ma'dan or Marsh Arabs. Made famous by the tales of the English explorer Wilfred Thesiger, the Ma'dan have lived in these marshes for over 5,000 years, largely isolated from the rest of Iraq by the extensive wetlands. Their culture has changed little over hundreds of years, and the lakes and marshes of southern Iraq are central to their livelihoods. The expansive reedbeds, for example, provide the Ma'dan with their staple building material, from which they construct their boats and artificial island houses, while the lakes

SYRIA

Buhayrat ath Tharthar

Baghdad

I R A Q

S A U D I

- ◼ Ramsar Sites
- ◆ Parks and Reserves
- ◆ Ramsar Sites/ Parks and Reserves
- ▢ Water Bodies
- ▢ Freshwater Marsh/ Floodplains
- ▢ Estuaries/ Coastal Wetlands
- ▢ Mangrove
- ▽ Delta

0 km 500
0 miles 250

N

•Tehran

IRAN

AFGHANISTAN

Mesopotamia

Tigris

Euphrates

Zagros

Gavkhouni Lake and marshes of the lower Zaindeh Rud

Hamun-e-Puzak, south end
Hamun-e-Saberi and Hamun-e-Helmand

Al Basrah•

Shadegan Marshes and mudflats of
Khor-al Amaya and Khor Musa

KUWAIT

Lake Parishan
Al Kuwayt and Dasht-e-Arjan

Mountains

Neyriz Lakes and Kamjan Marshes

PAKISTAN

ARABIA

Khuran
Straits

Sheedvar Is.

Deltas of Rud-e-Shur,
Rud-e-Shirin and Rud-e-Mindab

Tubli Bay BAHRAIN
Al Manamah

Strait of
Hormuz OMAN

Deltas of Rud-e-Gaz and Rud-e-Hara

Hawar Islands

•Doha

QATAR

The Gulf

Govater Bay and Hur-e-Bahu

Ormara Turtle Beaches

Jiwani Coastal Wetland

Astola Is.
(Haft Talar)

Riyadh •

•Abu Dhabi

UNITED
ARAB
EMIRATES

OMAN

•Muscat

Gulf of Oman

on which the houses float have always been important fishing grounds.

While the Madan were traditionally spear-fishermen, catching species of barbel and carp for their own needs, they have begun in recent decades to use nets to improve their catch for export to Basra and Baghdad. Mat-weaving has also become an important source of income, as demand for the versatile coverings, which are used in packaging and fencing, as well as a building material, has increased. However, it is buffalo that have remained the basis of family wealth. As well as providing the Ma'dan with meat and hides, which are used to make a variety of products, the herds of buffalo also generate milk, which is turned into butter and yoghurt.

A threatened existence

Following the Gulf War of 1990/91, the Ma'dan suffered from continuing violence in the region. Their villages were attacked and bombed by helicopter gunships. This was followed by an aggressive drainage program that sought to reclaim land for agriculture and pursue rebels

179

hiding in the marshes. By 2003, 90% of the 7,725 square miles (20,000 square kilometers) had been lost, in what many regard as one of the world's most important environmental disasters. As a result the Ma'dan fled. Of the 200,000 estimated to live in the marshes in 1991 (down already from 400,000 in the 1950s) only 20,000 remained in 2003. Some 40,000 had moved to camps in Iran.

With the fall of Saddam Hussein in 2003, considerable international attention is now being focused on how to restore the marshes. Initially some of the earth and concrete dams built by Saddam were breached by the local people themselves, and this brought water to some areas in 2003 and 2004. However substantial international investment will be required if the marshes are to be restored to their original condition and extent.

Man-made wetlands

The rapid urban and agricultural development in the Arabian Peninsula in recent decades has resulted in the creation of a large number of artificial wetlands. The majority are either water storage reservoirs, areas of spillage from irrigation systems, sewage treatment ponds or artificial lagoons created by waste water from urban and industrial areas. Some of these wetlands can be surprisingly large, and many rank among the most

▼ *Landsat satellite imagery reveals that in the last decade, wetlands that once covered as much as 7,725 sq mi (20,000 sq km) in parts of Iraq and Iran have been reduced to about 15 percent of their original size. Through the damming and siphoning off of waters from the Tigris and Euphrates Rivers, the ecosystem has been decimated and, as a result, a number of plant and animal species face possible extinction. In the first image, dense marsh vegetation appears as dark red patches. The second image shows the state of the*

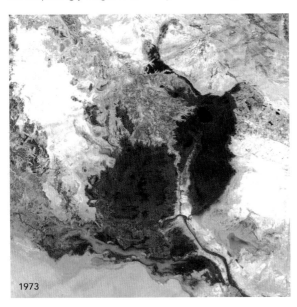

1973

1990

marshlands in 1990, shortly after the Iran-Iraq war. The scene reveals that a large eastern swath of the Central and Al Hammar Marshes as well as the northwestern and southern fringes of the Al Hawizeh Marsh had dried out by this point. In the final image most of the Central Marshes appear as olive to grayish-brown patches indicating low vegetation cover on moist to dry ground. The very light to gray patches are areas of exposed ground with no vegetation, which may actually be salt flats where before there were lakes.

important freshwater wetland habitats for wildlife in the peninsula.

A number of the larger artificial wetlands, such as the reservoirs in Wadi Hanifah and Wadi Jizan in Saudi Arabia, support several thousand waterfowl during the winter months. In addition, sewage lagoons and irrigation ponds provide excellent stop-over areas for a wide variety of migrant birds.

Azraq Oasis

The wetlands of Azraq Oasis, in Jordan's Eastern Desert, lie at the heart of a large internal drainage basin. The basin extends from the Druze Highlands in Syria to the borders of Saudi Arabia. The oasis wetlands themselves formerly comprised about 6 square miles (15 square kilometers) of permanent areas of marshes and pools, which lay alongside a seasonally flooded *playa* wetland known as Qa Azraq.

Originally covering an area of about 23 square miles (60 square kilometers) Qa Azraq was, and indeed still is, an outstanding example of a seasonal wetland in a predominantly arid region. As well as supporting a rich and varied flora and fauna, the wetland is a vitally important "refueling" area for huge numbers of migratory birds. Qa Azraq was designated as a Ramsar Site in 1977.

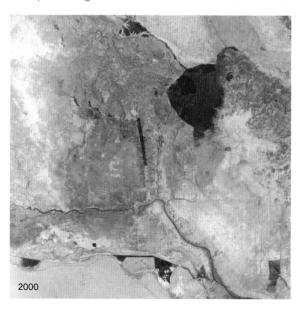

2000

A stretched resource

Since 1982, however, the spring-fed marshes have suffered drastically as a result of wide-scale groundwater extraction. The amount of water pumped to Amman, for example, was increased by 60 percent, while there has also been a rapid proliferation in the number of wells extracting water for irrigation. By 1991, there were some 450 irrigation wells, mostly illegal, pumping an estimated 790 million cubic feet (22 million cubic meters) of groundwater per year, and the water table had fallen by between 20 and 33 feet (6 and 10 meters). By the end of 1992 the four main springs which feed the wetland had dried up.

In response to these problems a major effort was made over the 1990s to protect Azraq. With support from the Global Environmental Facility an additional water supply was provided to the wetland area, but this is likely only to be a short-term solution. For the long term the concern continues that one of the world's most

▲ *The pools and marshes at Azraq Oasis in Jordan represent the water-wealth of a vast arid area. In recent years, however, extraction of groundwater has lowered the water table, and the wetlands are now vastly reduced.*

outstanding oasis wetlands, which was formerly capable of supporting more than 100,000 wintering waterfowl and perhaps as many as a million migrant birds, will never return to its former condition.

The Sarmatic Fauna

Ten of the waterbirds most closely associated with the Middle East are believed to have evolved around the shores of the ancient Sarmatic Sea. About three million years ago, this vast saline inland sea extended as a continuation of the present eastern Mediterranean, through the Black Sea and Caspian Sea to the Aral Sea and beyond. As this sea became fragmented into a chain of smaller inland seas and salt lakes, the populations of waterbirds associated with it similarly became fragmented, and most of these species are now extremely scarce.

While a few of the species, such as the red-crested pochard (*Netta rufina*), great black-headed gull (*Larus ichthyaetus*) and slender-billed gull (*L. genei*) remain at least locally common, four species, namely the Dalmatian pelican (*Pelicanus crispus*), pygmy cormorant (*Phalacrocorax pygmeus*), marbled teal (*Marmaronetta angustirostris*) and white-headed duck (*Oxyura leucocephala*) are now considered at risk.

Conservation in the Middle East

Because of the significance of wetlands in arid regions much greater investment in wetland conservation are required in the Middle East. At present Iran leads the way in efforts to conserve wetland resources. This befits the country where the Convention on Wetlands of International Importance (the Ramsar Convention) was drawn up – in 1971 in the town of Ramsar on the shores of the Caspian Sea.

Iran's first protected area to incorporate a major wetland (Lake Orumiyeh) was established in 1967 by the Iran Game and Fish Department, later to become the Department of the Environment. By 1977, this department had established no less than 68 protected areas and refuges, covering over 30,000 square miles (78,000 square kilometers). Within this area, examples of all the country's major natural ecosystems were represented. This excellent system of reserves survived the political upheavals of the late 1970s almost intact, and today 22 sites are listed under the Ramsar Convention.

East Africa and the Nile Basin

The landscape of East Africa is dominated by three physical features: the Great Rift Valley, which runs through the eastern side of the continent, Lake Victoria, and the valley of the River Nile as it flows north to the Mediterranean. These physical features, together with the coastal systems of the Indian Ocean and the Red Sea support a variety of wetlands ranging from the vast expanse of the Sudd floodplain in southern Sudan to the glacial lakes of Mount Kilimanjaro.

The high-rainfall areas of East Africa surround Lake Victoria and provide the Nile with more than half its water. The entire surface of Uganda drains its waters into the Nile Basin, as do parts of Kenya, Tanzania, Rwanda and Burundi. This wet region supports several large and many small wetlands associated with its rivers and lakes, the largest being the edges of Lake Victoria, which extend through three countries. The more moderate rainfall of Tanzania feeds several large floodplain wetlands, the most notable being the Malagarasi/Moyowosi system, which drains into Lake Tanganyika, the Kilombero floodplain, which drains into the Rufiji River and Indian Ocean, and the Wembere floodplain, which drains into the Rift Valley.

- Water Bodies
- Freshwater Marsh, Floodplains
- Mangroves
- Swamp, Flooded Forest
- Salt Pans
- Alkaline/Saline Lake

▼ *The Rusizi river flowing into lake Tanganyika. The delta of the Rusizi forms an important wetland and provides habitat for resident and migratory waterbirds. The area is also used intensively by local people, many of whom have come to the lake shore to escape fighting in the great lakes region.*

0 km 500
0 miles 250

N

E G Y P T

Buheirat en Naser

*R e d
S e a*

Nile

'Atbara

Khartoum •

Asmara •
ERITREA

S U D A N

White Nile

Blue Nile

Lake
Tana

E t h i o p i a n

H i g h l a n d s

Bahr el Ghazâl

Addis Abeba •

S û d d

ETHIOPIA

Lake Zwai
Lake Langana
Lake Shala

Bahr el Jebel

CENTRAL
AFRICAN
REPUBLIC

GREAT

RIFT

Lake
Abaya

VALLEY

Juba •

Lake
Chamo

*Lotagipi
Swamp*

UGANDA

Lake
Turkana

K E N Y A

Most of neighboring Kenya is much drier. Its large mountains, Mount Kenya and Mount Elgon, give rise to rivers that cross the desert and develop important floodplain wetlands en route – especially the Tana River floodplain and delta. Several seasonal floodplains and swamps are present in dry northern Kenya, the two most important being the Lorian in the northeast and the

Lotagipi in the northwest, which is fed by watersheds in neighboring Sudan. The highest mountain in the region is Mount Kilimanjaro in Tanzania, which despite being so close to the Equator, has glaciers and melting snow that feed springs further down its slopes, producing swamps and floodplains.

The Nile and Horn of Africa

As the Nile flows north into the Sudan, it spreads out to form the Sudd floodplain. Covering 6,400 square miles (16,500 square kilometers) of permanent swamp and up to 5,800 square miles (15,000 square kilometers) of seasonal floodplain, the Sudd is one of the three largest wetlands in Africa. As it leaves the Sudd, the White Nile is joined by the Blue Nile, which drains the Ethiopian Highlands. In this mountainous country there is a diversity of wetlands ranging from the extensive salt marshes and isolated mangroves of the Red Sea coast and the lakes of the Rift Valley, to the montane lakes and seasonal floodplains of the rivers draining the mountains. In Somalia, wetlands are confined largely to a few lagoons, mangroves and other tidal wetlands on the coast, and

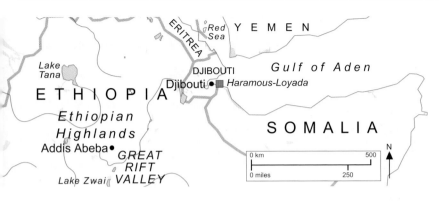

Ramsar Sites
Water Bodies
Freshwater Marsh, Floodplains
Tidal/ Coastal Wetlands
Mangroves
Swamp, Flooded Forest
Salt Pans
Alkaline/Saline Lake

swamps and floodplains along the Jubba and Shebelle, the two large perennial rivers which cross the country from Ethiopia. Similarly in Eritrea and Djibouti there are only a few coastal wetlands and small reservoirs.

In Egypt, the Nile today is tightly constrained, first by the Aswan Dam and by canals, dikes and irrigation structures along its course to the Mediterranean. These changes have reduced the delta and floodplains of the lower Nile to a remnant of their former selves. The vast papyrus swamps, for example, which were characteristic of the delta during the ancient Egyptian civilization, have almost totally disappeared from the region.

Coastal wetlands

The main coastal wetlands lie in deltas and estuaries where extensive stands of mangroves have developed, particularly in the Tana Delta and Vanga-Funzi in Kenya, and the Pangani and Rufiji deltas in Tanzania. The mangroves are currently threatened by overharvesting for construction poles, firewood and charcoal, and timber for furniture. Clear-cutting for urban development, creation of prawn farms, expansion of port facilities, diversion of freshwater, oyster harvesting, and siltation from the upper catchment areas have also been responsible for their degradation.

Lakes of the Rift Valley

Stretching from Lebanon to Mozambique, the Great Eastern Rift Valley is a 4,000-mile (6,500-kilometer) fissure in the Earth's crust. In East Africa, it is at its most dramatic and provides the foundation for a series of lakes which are among the most exotic features of the region's

◄ Cattle drinking from the river bank, Ethiopia. Livestock play a critically important role in the rural economy of Ethiopia. Rivers and wetlands provide an important source of water and pasture for grazing.

wetland diversity. There are three major types of lakes: freshwater, moderately saline and hypersaline.

In Ethiopia, there are six lakes strung along the Rift Valley – Zwai, Langana, Abiata, Shala, Abaya and Chamo. In Kenya, the line is picked up and continued southward through a chain beginning with the biggest and moderately saline Lake Turkana, then continuing along the valley with lakes Baringo, Bogoria, Nakuru, Elmenteita, Naivasha and Magadi. Again, these lakes vary from freshwater to hypersaline. In Tanzania, the chain continues with lakes Natron, Manyara, Eyasi and Rukwa; while in Malawi, Lake Malawi also lies in the Rift Valley.

The western arm of the Rift Valley includes lakes Kivu, Edward, Albert and Tanganyika. Lake Tanganyika, with a maximum depth of 4,850 feet (1,470 meters), is the second deepest lake in the world after Lake Baikal in Siberia. The lake is home to over 500 animals that are unique to the area, most of which evolved within the lake basin itself.

The most serious immediate problems facing the Rift Valley lakes have come about from population growth within the lakes' basins. Excessive suspended sediment inputs into the lakes caused by deforestation and overgrazing, pollution, eutrophication, introduction of exotic species and overharvesting of the fragile aquatic resources have resulted in serious reductions of species and localized extinctions.

▦	Ramsar Sites
◆	Ramsar Sites/ Parks and Reserves
▢	Water Bodies
▢	Freshwater Marsh, Floodplains
▢	Tidal/ Coastal Wetlands
▢	Mangroves
▢	Peatlands
▢	Swamp, Flooded Forest
▢	Salt Pans
▢	Alkaline/Saline Lake
▽	Delta

Soda lakes

Several of the lakes of the Rift Valley have exceptionally high concentrations of salts, and are known as soda lakes. Their waters are usually thick, slimy and greenish-looking because of the dense population of phytoplankton. Fish, however, are absent from most of the soda lakes, although one small species, *Tilapia alcalica grahami*, is able to tolerate the highly alkaline water of Lake Magadi in Kenya. This species was also introduced to Lake Nakuru (also in Kenya) with great success in 1961, and has encouraged increasing numbers and varieties of waterbird to the lake to feed.

Soda lakes are also famous for their large concentrations of birdlife, and some such as Lake Nakuru, which sup-

■ Montane Wetlands

High in the Ruwenzori Mountains and on the slopes of Mount Kenya are small alpine lakes formed by glaciation. These alpine lakes are small, clear, cold and sometimes frozen. There are 32 such lakes on Mount Kenya alone, with the highest, Harris Tarn, lying at some 15,800 ft (4,800 m). They are bordered by the montane vegetation zone peculiar to the eastern Africa region and dominated by giant lobelias (genus Lobelias) and groundsels (genus Senecio). Many of the lakes derive a greenish color from finely divided glacial silt, giving rise to names such as Emerald Tarn, which denotes the green glacial nature of their waters.

ports over a million lesser flamingos at any one time, have been declared national parks.

Tana Delta wetlands

The Tana River Delta covers an area of approximately 500 square miles (1,300 square kilometers) and is the largest delta ecosystem in Kenya. Its riverine forests

support two critically endangered and endemic species of primates, the Tana River red colobus monkey (*Procolobus rufomitratus*) and the Tana River crested mangabey (*Cercocebus galeritus*), while the 260 square miles (670 square kilometers) of floodplain grasslands, and associated woodlands and bushlands, lakes, mangroves, sand dunes and coastal waters, form a complex system which is unique on the Indian Ocean coast of Africa. Traditional land use practices of small-scale agriculture, pastoralism and fishing have maintained the ecological balance of the delta for thousands of years. The delta provides dry-season grazing and during droughts supports about 300,000 cattle. It also maintains large numbers of wild herbivores; up to 10,000 topi (*Damaliscus lunatus*), thousands of waterbucks (*Kobus ellipsiprymnus*), hippos and about 100 elephants.

In Kenya's dry coastal province the water of the Tana has always been used for a variety of purposes, and there have been a range of projects that have sought to improve agricultural production by harnessing the river. Three such projects collapsed in the 1980s but a new rice scheme was developed in the 1990s. The long-term impact of these projects on the delta is a major concern

▲ *Lake Nakuru, Kenya is internationally famous for the large numbers of flamingos that occur there. Over one million lesser flamingos have been observed at lake Nakuru.*

for the people who are dependent on other resources of the delta.

For most of the past 20 years the international image of southern Sudan has been one of drought and civil war. Every year, the world's media has brought us images of refugees fleeing by the thousand and famine on a massive scale. Yet in the midst of this human disaster lies the largest wetland system in East Africa. The Sudd, the Arabic word for "obstacle", is a vast floodplain where the White Nile leaves Uganda and flows north. Beginning 100 miles (160 kilometers) from Juba, and extending for some 340 miles (540 kilometers) to Malakal, the Sudd is a maze of thick, emergent aquatic vegetation. This vast wetland system is home for the Dinka, Nuer and Shilluk tribes, who have adapted their lives to the seasonal cycle of flood and drought.

When floods reach a peak in the latter half of the wet season, people congregate on high ground, bringing their cattle to graze on the upland grasslands that flourish with the July to October rains. At the same time, fields can be cultivated, weeded and protected from pests, and cattle herded in ever-increasing proximity as the countryside is transformed from an arid plain into a shallow inland sea. When the floods recede, the cattle are herded onto the floodplain or "toic". Here, seasonal inundation produces a rich cover of aquatic grasses, which remain for the dry season. Wildlife also concentrates on the floodplain during this period and hunting provides an important source of food.

Beyond the floodplain lies the permanently flooded swamp characterized by papyrus, cattails and open water. Here, specialist fishing communities live in villages on the banks of the main river channel, or on mounds that have been built up from detritus over the years. These people subsist largely on fish and what other foodstuffs they can trade for fish.

The Jonglei Canal

As the Nile flows through the Sudd the volume of water is reduced through evaporation and evapotranspiration. For much of the last century considerable national and international attention was focused on how to reduce this "loss" of water. In response construction of the Jonglei canal was started in 1978. If completed, this canal would stretch 210 miles (360 kilometers) to bypass the Sudd and direct downstream some of the water that now floods the swamp. However, it is these "losses" which help yield the pasture, crops and fishery resources that are the basis of the local economy. The proposed diversion of the river has given rise to concern that the canal will destroy the livelihood of the peoples of the Sudd.

In the middle of much international attention, work on the Jonglei Canal started in 1978. However, the start of the civil war in 1983 brought all canal-related development to a stop. With a peace agreement now in place and attention focusing upon how to meet the development needs of the region, a careful reappraisal of the Jonglei project and its effects on the people of the Sudd will be necessary.

Uganda's response

For centuries, Uganda's wetlands have been important resources for the country's inhabitants. Today, with rising population and increased pressure upon dryland resources, Uganda's wetlands are of even greater significance. At least 10 percent of Uganda is made up of swamps and floodplains, and the edges of the main lakes – Victoria, Kyoga, Albert, Edward and George – are of prime importance. Wetlands are, therefore, a national feature; they provide food, building materials (in the form of reeds) and water.

Since the latter half of the 1900s, Uganda's wetlands have come under increasing pressure from drainage for extension of pasture and crops, for rice paddy development or for expansion of dry land for other development projects. Agricultural and industrial pollution, watershed

▼ *Swamplands are particularly common in the Nile drainage basin where immense papyrus swamps occupy several thousand square miles of southern Sudan and Uganda.*

1 Matetite reed
  Phragmites sp.
2 Hippopotamus
  Hippopotamus amphibius
3 Hammerhead stork
  Scopus umbretta
4 Black crake
  Limnocorax flaviostra
5 Sitatunga
  Tragelaphus spekki
6 Swamp worm
  Alma emini
7 Saddlebilled stork
  Ephippiorhynchus senegalensis
8 Water cabbage
  Pistia stratiotes
9 Water lily
  Nymphaea sp.
10 Bichir
  Polypterus sp.
11 Catfish
  Malapterurus sp.
12 Lily trotter
  Actophilornis africana
13 African spoonbill
  Platalea alba
14 Papyrus
  Cyperus papyrus
15 Malachite kingfisher
  Corythornis cristata
16 Herald snake
  Crotaphopeltis hotamboeia
17 Shoebill
  Balaeniceps rex
18 Squacco heron
  Ardeola ralloides
19 Snail
  Biomphalaria sudanica
20 Marsh mongoose
  Atilax paludinosus
21 Lungfish
  Protopterus aethiopicus

degradation, land use and land tenure disputes, invasion of aquatic weeds and overuse of resources have also caused wetland loss. Swamp products such as papyrus have also been overharvested, vegetation overgrazed, and the capacity of some wetlands to purify domestic and urban sewage, and agricultural run-off has become overstretched.

Managing Uganda's wetlands

In response to the pressures facing Uganda's wetlands the country has developed one of the world's most advanced approaches to wetlands conservation and management. It has to be, since wetlands represent one of the country's most vital economic resources; the services and products provided by wetlands contribute several hundred million US$ to the Ugandan economy each year.

The importance of wetlands for Uganda was given formal recognition in 1986 when the Government placed a ban on further wetland conversion until a National Wetlands Policy was developed. Policy development started in 1989 and from this process a fully-fledged National Wetlands Program evolved to begin the task of policy implementation. Fifteen years later, wetlands are now placed firmly on the country's political and public agenda, and considerable progress has been made toward developing and applying the tools needed for sustainable wetland management through strengthening legislation and institutional structures, and through capacity building and awareness-raising at all levels.

Uganda's National Wetlands Policy was the first of its kind in Africa. Launched in 1995 the policy is implemented at national, district and local levels and is supported by a dedicated government department, the Wetlands Inspection Division (WID) in the Ministry of Water Lands and Environment. The establishment of WID in 1998 was a breakthrough in wetland management in Uganda and unprecedented in Africa.

By conducting a national wetlands inventory the government has also started classifying and prioritising wetland for management purposes. For vital wetlands whose services and products cannot be replaced by any other means, procedures are being developed to give them strict protection. For other wetlands management systems that allow for sustainable use by communities are being developed.

West and Central Africa

West and Central Africa is a region of contrasts. It stretches from the peninsula of Cap Vert in Senegal to eastern Chad, from the northern border of Mali to the Nigerian coast, and from the mountains of northern Chad to the Congo basin. Travelers in the 1700s found that the region contained some of the most spectacular and varied landscapes in Africa. Within the area, for example, lie the deserts of Mauritania, Mali and Niger, the Sahelian "desert edge" and the savanna and extensive tropical forest of the south.

Although population increase and other changes over the last 100 years have caused substantial loss and degradation of natural ecosystems, many still remain intact today and are of central importance in the lives of the people of the region. Wetlands are among the most important of these ecosystems, and remain central to the regional economy.

Floodplains and flooding

It is one of the ironies of the African landscape that three of the continent's most important wetlands, namely the floodplains of the Senegal, Niger and Chad river basins, are found in the semiarid region of the Sahel. The Senegal and Niger rivers rise in the Fouta Djallon massif of Guinea, where the wet season can last for seven to nine months and annual rainfall can reach 120–200 inches (3,000–5,000 millimeters). The rivers then continue north through the wooded savanna and into the Sahel where, with a wet season of only three to four months,

◄ Coastal marshes, Guinea Bissau. The country's sharply indented coastline supports an important diversity of wetlands, including tidal flats, mangroves, and fresh and brackish water marshes. During the European winter these wetlands support internationally important populations of migratory shorebirds.

WESTERN SAHARA

Sahara

ATLANTIC OCEAN

MAURITANIA

Parc National du Banc d'Arguin

• Nouakachott

Lake d'Aleg
Lake Riz
Chat Tboul
Parc National du Diawling
Djoudj
Bassin du Ndiaël
Gueumbeul

Sénégal

Dakar • SENEGAL

M A L I

Delta du Saloum
Baobolon Wetland Reserve
Banjul • GAMBIA
Gambia

Bafing

Niger

• Bamako

Bissau •
GUINEA-
BISSAU
Lagoa de Cufada
Iles Tristao
Ile Alcatraz
Rio Kapatchez
Rio Pongo
Konkouré
Ile Blanche • Conakry
GUINEA
Tinkisso
Niger-Tinkisso
Niger-Niandan-Milo
Sankarani-Fié
Niger-Mafou
Niger Source

Sierra Leone River Estuary
Freetown
SIERRA LEONE

IVORY COAST

Lake Piso
Monrovia •
LIBERIA

Ramsar Sites
Ramsar Sites/
Parks and Reserves
Water Bodies
Freshwater Marsh/
Floodplains
Tidal/Coastal Wetlands
Mangroves
Swamp/Flooded Forest
Salt Pans
Delta

N

0 km   500
0 miles   250

195

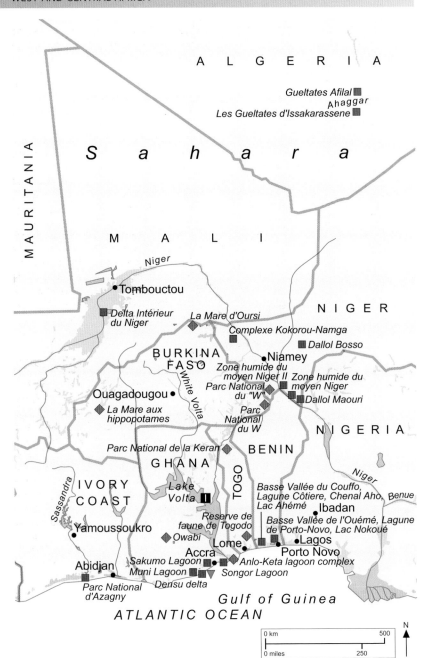

ALGERIA

Gueltates Afilal ■
Ahaggar
Les Gueltates d'Issakarassene ■

S a h a r a

MAURITANIA

M A L I

Niger

● Tombouctou

■ Delta Intérieur   La Mare d'Oursi
du Niger

NIGER

◆ Complexe Kokorou-Namga
■ Dallol Bosso

BURKINA
FASO   ● Niamey

Zone humide du
moyen Niger II   Zone humide du
Parc National ◆   moyen Niger
du "W" ■   ■ Dallol Maouri

Ouagadougou ●

White Volta

◆ La Mare aux   Parc
hippopotames   National ◆
du W

NIGERIA

Parc National de la Keran ◆   BENIN

GHANA   Niger

IVORY   Lake   TOGO   Basse Vallée du Couffo,
COAST   Volta ■   Lagune Côtiere, Chenal Aho, Benue
                                       Lac Ahémé   ● Ibadan
Sassandra   Reserve de
                        faune de Togodo   Basse Vallée de l'Ouémé, Lagune
● Yamoussoukro   ◆ Owabi   de Porto-Novo, Lac Nokoué

                  ● Accra Lome ●   ● Lagos
Abidjan   Sakumo Lagoon ■ ■ ◆ Anlo-Keta lagoon complex   ● Lagos
●   Muni Lagoon ■ ▼ Songor Lagoon   Porto Novo
■   Densu delta
Parc National
d'Azagny   Gulf of Guinea
ATLANTIC OCEAN

0 km   500
|----------|----------|----------|----------|
0 miles   250

N ▲

▌ Lake Volta

Lake Volta is the largest artificial lake in Africa. Created in 1964 by the closure of the Akosombo Dam 60 miles (100 km) upstream from the mouth of the Volta, the reservoir covers 330,000 sq miles (850,000 sq km) and stretches 190 miles (320 km) from north to south and 250 miles (400 km) from east to west. Built to provide hydroelectricity, the dam has rarely achieved full capacity as it takes 3 or 4 years of consecutive flood to fill the lake to its full capacity. The construction of the dam led to a substantial reduction in the fisheries of the Volta Delta.

lakes and temporary marshes are the only wetlands of any significance beyond the floodplains. The area of each floodplain varies from month to month as the rivers flood and recede, and from year to year in accordance with rainfall variation in the catchment area.

Flooding is the driving force behind the productivity of the floodplains and of central importance to the role which wetlands play in the region. From as early as the 10th century, these floodplains have been centers of population in an otherwise arid and thinly populated region. Today, the pastoral and agricultural economy of much of the Sahelian region is dependent upon the presence of these floodplains to provide dry-season grazing, a regular supply of freshwater for agriculture, and the principal source of animal protein, in the form of fish, for much of the year.

Coastal wetlands

Along the coast, rivers have led to the development of several extensive wetland systems over hundreds of thousands of years. Mauritania's coast is dominated by the Banc d'Arguin, a former estuary which today forms Africa's largest coastal wetland. To the south, the coastal deltas of the Senegal, Saloum, Gambia, Casamance and Niger rivers add to the extent and diversity of these wetlands. The Bijagos Archipelago, which millions of years ago was once the delta of the Rio Geba and the Rio Grande de Buba in Guinea-Bissau, complements the Banc d'Arguin with its mosaic of palm-covered islands and extensive areas of mangrove.

The mangroves form part of a band of 3,300 square miles (8,500 square kilometers) from Senegal to Sierra Leone, while the coastal delta of the Niger in Nigeria supports an additional 4,000 square miles (10,000 square kilometers) of mangroves. Together with the coastal lagoons of Ghana, the Ivory Coast, Benin and Togo, the Niger Delta dominates the Gulf of Guinea and plays a critical role in supporting the region's rich wildlife, which includes manatees, pygmy hippos (*Choeropsis liberiensis*), three species of crocodile, turtle, forest elephants and chimpanzee.

▦ Ramsar Sites

◈ Ramsar Sites/
Parks and Reserves

▢ Water Bodies

▢ Freshwater Marsh/
Floodplains

▢ Tidal/Coastal Wetlands

▢ Mangroves

▢ Swamp/Flooded Forest

▢ Salt Pans

▽ Delta

▲ *Pelicans at Diawling National Park, Mauritania. Lying on the right (northwestern) bank of the Senegal river, the Diawling is on e of Mauritania's three Ramsar sites. With construction of the Diama dam on the Senegal river in 1986, the Diawling was cut off from the annual river flood. This has now been restored through a system of sluice gates that allow water to be released from the Diama reservoir.*

The Congo River system

The most extensive wetlands of the region extend across the basin of the Congo river. These cover over 80,000 square miles (200,000 square kilometers) of the Congo basin, and are composed of swamp forest interspersed with islands and peaty hummocks. The scale of this vast wetland system reflects the almost year-round flow of the rivers of the basin, the largest of which is the Congo river itself. In places the floodplain forest and grasslands extend as far as 3 miles (5 kilometers) from the shore, and are very sparsely populated. Hunters and fishing communities live along the major rivers, which also provide a crucially important means of transportatioin and communication in a region with few roads. The soils have limited agricultural potential and the flooded landscape prevents commercial exploitation of the timber resources.

Within the Congo basin lies the Salonga National Park. Covering 14,000 square miles (36,000 square kilometers), the park is one of the world's largest forest reserves. The park comprises a large area of swamp forest which provides an important reserve for the endemic pigmy chimpanzee (*Pan panicus*) and the Congo "peacock" (*Afropavo congensis*). A number of pygmy tribes also live in the park, basing their subsistence economy on the many and varied resources of the forest.

The Banc d'Arguin

The Banc d'Arguin is the largest intertidal wetland

system in Africa. Lying between the two headlands of Cap Blanc to the north and Cap Timiris to the south, the Banc d'Arguin is a vast expanse of intertidal flats and creeks, where beds of eel grass (*Zostera and Cymodocea* sp.) and other habitats provide important breeding and nursery areas for fish and crustaceans.

In the Baie d'Arguin, and surrounding the island of Tidra, two archipelagos of sand and rock (sandstone) islands support large breeding colonies of 15 species of waterbird. Found among the 25,000–40,000 pairs, there are great white pelicans (*Pelicanus onocrotalus*), greater flamingos, an endemic subspecies of the European spoonbill (*Platalea leucorodia balsacii*), and several species of heron, egret and tern. During the European winter, the tidal flats surrounding these islands provide a rich feeding ground for more than 2.3 million shorebirds, including bar-tailed godwit (*Limosa lapponica*), dunlin (*Calidris alpina*) and both ringed and gray plovers (*Charadrius hiaticula* and *Pluvialis squatarola*).

In addition to this avian diversity, the combined influence of the cold Canaries current from the north and the warm Guinean current from the south make the Banc d'Arguin a frontier zone. Here, many plant and animal species from northern Europe and Asia at the southern limit of their range mingle with Afrotropical species at their northern limit. For example, the mangrove

▼ *The Porto Novo Lagoon Ramsar site, near Cotonou, Benin. Throughout West Africa, coastal wetlands are used intensively by local people for fishing. Some 21,000 tons of fish, crabs and shrimp are produced from the Porto Novo Lagoon system each year. This employs some 24,000 fishermen and 13,000 seasonal workers, and in total supports the livelihoods of some 200,000 people.*

ALGERIA   LIBYA

Gueltates Afilala
Ahaggar
Les Gueltates d'Issakarassene

S a h a r a

N I G E R

C H A D

Lac Tchad   Partie tchadienne du lac Tchad

Lake
Chad   Lac Fitri

•N'Djamena

Chari

Logone

Nguru Lake (and Marma
Channel) complex

N I G E R I A

Benue

Niger

CENTRAL
AFRICAN REPUBLIC

Ubangi

C A M E R O O N

Sanaga

•Douala   Bangui

Malabo•   DEM. REP.
OF CONGO

Gulf of   •Yaounde   CONGO
Guinea   EQUATORIAL
GUINEA

Avicennia germinans is the most northerly in Africa, while the grass *Spartina maritima* is at its most southerly limit. Similarly, while the monk seal (*Monachus monachus*) and the common porpoise (*Phocoena phocoena*) are at the southerly limit of their range, the Atlantic humpback dolphin (*Souza teuzsii*) is not found north of the Banc d'Arguin.

The rich faunal and floral diversity of the region prompted the establishment of the Banc d'Arguin National Park in 1976, and its subsequent inclusion in the Ramsar List of Wetlands of International Importance and listing as a World Heritage Site. In 2000 the Banc d'Arguin was listed as a WWF Gift to the Earth.

The Imraguen fishermen

Every November to January, the Imraguen people fish for migrating mullet in the shallow waters of the Banc d'Arguin. Using long poles to beat on the water, they attract dolphin, which drive the mullet inshore and into their nets. The Imraguen's close dependence upon the tidal system, and their carefully adjusted fishing techniques, have made them the principal guardians of the Banc d'Arguin in the face of increasing fishing pressure from foreign fishermen. At a time when revenue from offshore fishing is Mauritania's principal source of export earnings, the Imraguen help protect the most important fish nursery in the country.

In 1986, the International Banc d'Arguin Foundation was created, the purpose of which was to assist in providing international support to the Banc d'Arguin National Park. Over the past 20 years, the foundation has supported the Banc d'Arguin National Park in working with the Imraguen to sustain their fishing livelihoods and the role

▲ *Dunlin* (Calidris alpina) *in fall (top) and winter (bottom).*

Ramsar Sites

Ramsar Sites/ Parks and Reserves

Water Bodies

Freshwater Marsh/ Floodplains

Tidal/Coastal Wetlands

Mangroves

Peatlands

Swamp/Flooded Forest

Salt Pans

Pools

2 Lake Chad

Spreading across the borders of Niger, Nigeria, Cameroon and Chad, Lake Chad, rarely more than 23 ft (7 m) deep, can cover 10,000 sq miles (25,000 sq km) when river flow is high, but recede to less than 4,000 sq miles (10,000 sq km) in times of drought. Ninety-five percent of the lake's water comes from the Chari and Logone rivers, which rise in the mountains of southern Chad, Cameroon and Central Africa. The remaining 5 percent come from diverse, intermittent rivers, in particular the El Beid, the Komadougou-Yobe, and the Yetseram, which rise in Nigeria.

they play as custodians of the Park. This faces many challenges, in particular the pressure placed upon the inshore fisheries by the offshore fleet and illegal fishing by the increasing number of fishing boats from further south in Mauritania and Senegal. For example the high value and export of the roe of Yellow Mullet (called poutargue) has led to such a rapid decline in stocks of this species that the Imraguen have greatly reduced their harvest. This is a monumental change for a society where mullet has played such a strong economic and cultural role for several centuries. To help the Imraguen adjust to these changes, and to prevent them being forced to exploit other resources unsustainably in order to maintain their livelihoods, the park has worked with international partners to develop alternative forms of sustainable fishing.

In the face of this progress however, new threats face the Banc d'Arguin. Oil has been discovered offshore and commercial exploitation will begin at the end of 2005. There are concerns about the potential impact of this industrial development on marine biodiversity and the long-term sustainability of the region's fisheries.

Ramsar Sites

Ramsar Sites/
Parks and Reserves

Water Bodies

Freshwater Marsh/
Floodplains

Tidal/Coastal Wetlands

Mangroves

Peatlands

Swamp/Flooded Forest

Bijagos Archipelago

Formed by the prehistoric delta of the Rio Grande de Buba and the Rio Geba, the Bijagos Archipelago consists of 88 islands and islets distributed over 4,000 square miles (10,000 square kilometers). Here, the rainy season brings freshwater into the coastal zone, while coastal currents from north and south meet, making the delta region vulnerable yet at the same time biologically rich. Between the islands, extensive mud flats are drained by a network of canals and creeks as the tide recedes.

Today, the characteristic vegetation of the islands are the palm groves which have replaced the once extensive forest. However, the tidal areas remain relatively untouched, forming a unique mosaic of mangroves and tidal flats. Here, hippos have adapted to life in sea water and can be seen plodding along the beaches, while otters hunt for shellfish or wallow in the creeks together with manatees for which the archipelago forms one of the most important strongholds in the region. Two species of dolphin also live here, including the rare Guinean dolphin (*Souza teuzsii*). Reptiles include two species of crocodile (*Crocodylus cataphractus* and *C. niloticus*) and four species of marine turtle, including the green turtle for which the Bijagos Archipelago is the

most important breeding site in West Africa. Some 7,000 female green turtles nest here each year, and tracking by satellite of turtles equipped with transmitters has shown that, after laying, the females move north along the coast toward the waters of the Gulf d'Arguin in Mauritania. This highlights the interconnectedness of these coastal wetlands and the importance of maintaining a network of sites if their biological diversity is to be preserved.

The people of the Bijagos

The archipelago is inhabited almost exclusively by the Bijagos ethnic group, 25,000 of whom live year round on 20 islands, with 20 others used at particular times of the year. The Bijagos assiduously maintain the palms from which they extract oil and palm wine; also harvesting shellfish from the mud flats and catching fish for daily consumption.

In pursuit of enhanced economic development, both government and private and foreign entrepreneurs have in recent years turned their attention to the archipelago. Increasingly, the region has become a focal point for development, with particular potential in tourism, and sport fishing. However, the fragility of the natural environment and the specific cultural heritage of the people of the Bijagos has given rise to concern that, unless it is carefully planned, the increased development investment will have serious social and environmental consequences. In addition excessive fishing pressure from both artisanal fishers from neighboring countries, and commercial offshore fleets threatens the sustainable of these fisheries. Some species, notably sharks, are particularly at threat due to the high value of their fins.

To address these concerns the government has prepared an integrated development plan for the islands. The archipelago has also been declared a Biosphere Reserve and this gives special protection to critical ecosystems while encouraging sustainable use of natural resources in other parts of the archipelago. Within the Biosphere Reserve there are two National Parks (João Vieira – Poilão; and Orango) and a further site (Urok Islands) is being established as a community protected area.

▼ *Fishing in Mali . The Inner Niger Delta is one of Africa's most important and spectacular wetlands. The local farmers, fishers and herders are intimately dependent for their livelihood on the annual flooding cycle which drives the productivity of the delta. Many hundreds of thousands of waterbirds use the delta as a feeding and nesting site.*

The Inner Niger Delta

During the 1400s, the Empire of Mali stretched from the coast of modern-day Senegal across southern Mauritania and Mali to the border of what is today Niger. To the south it encompassed the catchments of the Senegal and Niger rivers in Guinea, from where gold was traded across the Sahara. At the center of this empire lay the Inner Niger Delta, the largest floodplain in West Africa, and a stable source of food for the population. Today, the Inner Niger Delta continues to play a central role in the lives of the region's people.

When the river is in full flood, the delta and its associated wetlands extend over some 120,000 square miles (320,000 square kilometers). It forms an inland sea in an arid region, where rainfall in some parts can be as low as 8 inches (200 millimeters). Over half a million people depend upon the delta, exploiting its rich soils, pasture and fisheries. As the annual rains begin to fall at the end of the dry season in June and July, local herders move their cattle from the dry bed of the delta to the higher ground on its edges where the rain stimulates fresh pasture growth. Over the next few months the herders move between pastures, exchanging milk products for millet and other dryland cereals grown by local farmers.

By September, toward the end of the wet season, floodwaters from the highlands of neighboring Guinea transform the delta into a green oasis, enabling farmers to cultivate floating rice and other crops along the mosaic of channels and lakes that form the delta. Shortly afterward, the cattle are brought back into the delta, where they feed on the luxuriant growth of floating grass (*Echinocloa stagnina*) known locally as "bourgou". At this time, the delta can support over a million cattle and a million sheep and goats. The dry season is also the peak fishing season. In good years the delta can yield in excess of 100,000 tons of fish and several hundred thousand people depend on this resource for food and income.

Southern Africa

In a landscape of meadows and forest in northwest Zambia, the Central African Plateau gives rise to the Zambezi River. Flowing to the other side of the continental divide, the Luapula River drains to the Congo. These two river systems, the Zambezi and Congo, drive the freshwater wetlands over much of southern Africa – from the small freshwater marshes of their catchment areas, to the extensive swamps and floodplains of their valleys.

The Luapula drains the vast swamps and floodplains of the Bangweulu region and its associated lakes – Bangweulu, Mweru and Mweru-Wantipa in Zambia and the Democratic Republic of the Congo (DRC). Another tributary of the Congo, the Lualaba, on the other hand, forms a complex landscape of floodplains, swamps and lakes, including Lake Upemba, in southern DRC. The Zambezi River and its tributaries receive most of their runoff from Zambia and eastern Angola. They drain the great floodplains of western Zambia and the Kafue Flats, as well as the smaller floodplains of the middle Zambezi, Cuanda, Chobe/Linyanti and Luangwa rivers. For centuries, these floodplains have been crucial to the rural economies of the region.

Nowhere is the importance of these wetlands more evident than along the Zambezi, where many communities

- ▦ Ramsar Sites
- ◆ Ramsar Sites/ Parks and Reserves
- ▢ Water Bodies
- ▢ Freshwater Marsh/ Floodplains
- ▢ Tidal/Coastal Wetlands
- ▢ Mangroves
- ▢ Peatlands
- ▢ Swamp/Flooded Forest
- ▢ Salt Pans
- ▢ Dambo Regions

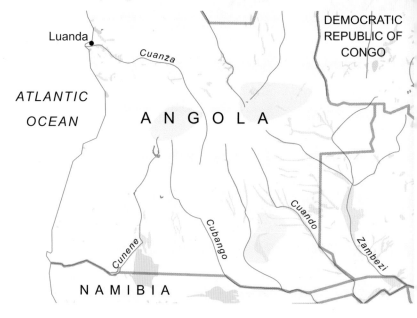

still organize themselves around the seasonal river flood. As the Zambezi flows back into Zambia from Angola it spreads out along the floodplains of Barotseland in what is today Zambia's Western Province. Stretching over 150 miles (250 kilometers) in length and 50–60 miles (80–100 kilometers) wide, the floodplain is flooded each year as the Zambezi rises above its low banks and spreads out over the landscape. As the grasses grow to keep above the flood, the plain becomes an intricate mosaic of green and blue.

■ Bangweulu Swamp

Lying at the center of the Bangweulu Basin in northeast Zambia, the Bangweulu wetland system includes several lakes and extensive swamps and floodplains, and is the largest and most diversified in Zambia. On the floodplain, black lechwe, tsessebe and sitatunga mix with flocks of ducks and geese. In 1991, a portion of the swamp was designated by Zambia under the Ramsar Convention.

The peoples of the floodplain

The Lozi people, who it is believed settled on the floodplain during the 1650s, continue to adapt to the annual flood cycle. Each year, as the river floods, the head of the tribe – the *litunga* – leads his people from the floodplain island that is their dry-season home to the higher ground along the edge of the valley where they spend the flood season. Today, this migration continues as a rich and colorful ceremony known as the *kuomboka*, marking an

important point in the year for the people of the flood-plain. Following nights of preparation, the chief's launch, crewed by 120 rowers, and a smaller barge for his queen, leads a flotilla of small boats across the lush greenery of the floodplain to the higher ground along the side of the valley.

While only a few of the larger wetlands match the grandeur of the Zambezi floodplain, this seasonal cycle is repeated across much of southern Africa, even in small freshwater marsh depressions, known as "dambos". Although tiny when compared to the extensive African floodplains, dambos, however, play a central role in agriculture and grazing in many countries. Much of the landscape of Malawi, Mozambique, Zambia, southern Angola, Zimbabwe, and parts of Tanzania is dotted by dambos. Dry dambos are characterized by rich aquatic grasses and are treeless, while wet dambos are occupied by evergreen forests. Dambos are especially important in

> **2 Etosha Pan**
> Covering 2,100 sq miles (5,500 sq km) of salt pans in northern Namibia, Etosha Pan is Namibia's most important wetland system. Although ephemeral, the pans when full are 6 ft (2 m) deep and then hold vast numbers of fish and waterbirds.

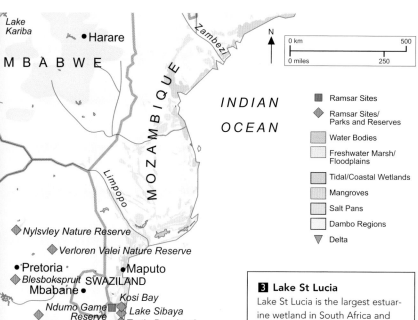

times of drought and during the dry season, when they frequently provide the only available grazing land, and provide the best crop harvests.

Lakes of southern Africa

The largest lake in the region, Lake Tanganyika, has limited wetlands because of its particularly steep basin and shores: it is an inland drainage system with a small outlet to the Congo River. Lake Malawi (Nyassa) is the other large natural lake with extensive wetlands on its edges and along the Shire River, which joins it to the Zambezi. Two large man-made lakes, Kariba and Cabora Bassa, impound the Zambezi and support viable wetlands when their management allows. In contrast to these large deep lakes,

3 Lake St Lucia

Lake St Lucia is the largest estuarine wetland in South Africa and together with associated swamps covers about 250 sq miles (660 sq km). With water entering the lake from four principal rivers and draining to the Indian Ocean through a 12-mile (20-km) estuary, the system combines a diversity of habitats, including mud flats, papyrus swamps, reed swamps, freshwater swamp forest, tidal swamp forest and freshwater pans and beaches. These support a diversity of flora and fauna; 129 species of invertebrate, 108 estuarine and 13 freshwater fish and 340 species of bird have been recorded for the system. St Lucia Game Reserve was established in 1897 and covers 140 sq miles (370 sq km) at the eastern shore and around the north and northwest of the lake, and in 1986 was listed under the Ramsar Convention.

lake Chilwa on the border of Malawi and Mozambique is relatively small and shallow, being less than 17 feet (5 meters) deep. At its maximum extent it can cover 1,500 square miles (2,400 square kilometers), but in some years it can dry out completely, the last being in 1996. Lake Chilwa is however highly productive and supports an important fishery that can yield up to 25,000 tons each year and in turn sustains the livelihoods of some 6,000 fishers. In 1996 Lake Chilwa was listed as Malawi's first Ramsar site.

Botswana and Namibia, the two driest countries in the region, have large and important seasonal and saline wetlands at Makgadikgadi and Etosha Pan. In northern Botswana, the Okavango Delta on the edge of the Kalahari Desert forms the most famous system of freshwater wetlands in southern Africa.

The most important coastal wetlands lie in the extensive mangroves of Mozambique and Angola, and the coastal lagoons and estuaries of Namibia and South Africa. The Zambezi Delta covers 500 square miles (1,300 square kilometers) and includes large areas of freshwater floodplain and mangrove forest. In Namibia, the sheltered bays of the Namib coast contain important coastal wetlands such as those at Sandwich Harbor and Walvis Bay.

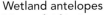 Ramsar Sites

Wetland antelopes
Many of Africa's wetlands are home to several of the continent's antelope species. The sitatunga (*Tragelaphus*

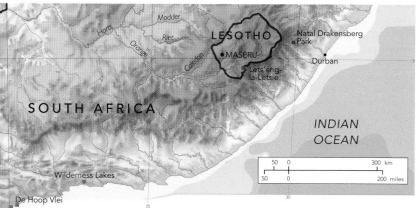

spekei) is the most closely adapted to this aquatic life and although largely confined to the deepest, "impenetrable" swamps, is widespread across the continent. More specific to the wetlands of southern Africa are the antelopes of the genus Kobus. Waterbuck (*Kobus ellipsiprymnus*) live in the riverine wetlands and floodplains of most areas. A special floodplain and dambo species in southern and eastern Africa is the puku (*K. vardoni*) from Malawi, Zambia, Namibia and Botswana. However, the best-adapted wetland kob of the great floodplains and swamps of southern Africa is the lechwe (*K. leche*).

Lechwe are medium-sized antelopes with long, lyre-shaped horns and elongated toes that help them to walk in their muddy environment. They feed on floodplain grasses both on the edges of floods and in the water, often wandering up to a depth of 3 feet (1 meter). When disturbed, lechwe will usually flee deeper into the flood-plains, using the water as a protection for their young, which are usually born on islands within the flooded grasslands or swamps. In the past, most populations of lechwe were prey to lions, which had learned to hunt them in the water. But in most places where lechwe live today, their main predator is man.

◀ *Plains zebra (Equus burchelli) at a water hole in Etosha National Park, Namibia. Etosha is one of Namibia's four Ramsar sites and is famous for its breeding population of lesser and greater flamingos (Phoeniconaias minor and Phoenicopterus ruber) during very wet years. At these times Etosha becomes one of southern Africa's great wildlife spectacles.*

Lechwe distribution

Three subspecies of lechwe inhabit the floodplains and seasonal wetlands of Democratic Republic of Congo, Zambia, Angola, Namibia and Botswana. In the past, the red lechwe (*Kobus leche leche*) was the most widespread. It ranged from southern Democratic Republic of Congo and northern Zambia, to Angola and Botswana.

Today, however, the red lechwe's most easterly home is the edge of Lukanga Swamp on the Kafue River in Zambia, where a small group shares its habitat with cattle. A protected population of red lechwe on Busanga Plain (in Zambia's Kafue National Park) numbers about 4,000 individuals, while scattered groups survive in the Barotse floodplains, the Cuando swamps and the Chobe/Linyanti floodplains close to the Zambezi. The largest group of red lechwe is found in the Okavango Delta of Botswana, where some 20,000 animals inhabit the seasonally flooded grasslands.

The largest single population of kob is the Kafue lechwe (*K.l. kafuensis*), of which as many as 50,000 use the Kafue Flats in Zambia. Further north, the black lechwe (*K.l. smithemani*) has a smaller, widespread population in the Bangweulu wetlands and, possibly, is still present in neighboring Democratic Republic of Congo.

▲ *Sitatunga*
(Tragelaphus spekii)

The Zambezi Basin

The Zambezi River and its major tributaries flow through six countries in central and southern Africa – Angola, Botswana, Mozambique, Namibia, Zambia and Zimbabwe – while both Malawi and Tanzania also supply water to its basin. All these countries wish to use the Zambezi's water, particularly for its agricultural and hydroelectric potential.

Yet rainfall, water and wetlands, as well as the human populations that seek to exploit these and other resources, are unevenly distributed throughout the basin. As a result, the dependence of people on the Zambezi varies. For example, although Zambia contributes at least half of the water flowing to the Zambezi, the country is relatively wet and rarely critically dependent upon the river's waters for its agriculture and domestic supply. In contrast, Botswana contributes little to the river's flow, but views it as a potential solution to the water shortage in the north of the country. This combination of distribution and demand is a source of potential conflict.

Wetlands and water demand

Many countries are currently facing water shortages and are looking to new sources for their supply. Among the potential candidates, the wetlands of the Okavango Delta, Chobe/Linyanti, Lake Kariba and Lake Cabora Bassa have all been considered as sources for "out-of-basin transfers" to alleviate water shortages – as has the flow of the Zambezi River itself. This has led to concerted efforts by scientists and conservation groups to demonstrate the economic value of these wetlands and of the water that sustains them. As a result there is now growing recognition that water needs to be retained in rivers and wetlands because of the role that these play in sustaining biodiversity and the rural livelihoods that depend upon this and the other environmental services provided by wetlands. In South Africa the new Water Law requires that a "water reserve" be maintained in rivers in order to meet the needs of ecosystems and the people who depend on these.

The Okavango Delta

Lying on the northern edge of the Kalahari Desert, the Okavango Delta has for centuries provided an oasis for the tribes of northern Botswana, and a refuge for some

▼ *The vast Okavango Delta stretches over some 11,000 sq miles (28,000 sq km). About half the land is permanent swamp, the rest is seasonal floodplain grassland, which is ideal for cattle ranching.*

of Africa's rarest birds and other wildlife. As pressure upon the region's water resources has increased, the future of the delta has become a major national and international concern.

Rising in the mountains of Angola, the Okavango River crosses into northern Botswana, where it flows for 60 miles (100 kilometers) along the "panhandle", a narrow valley between two geological fault lines, before spreading out to form the world's largest inland delta. From here to the downstream end of the delta, 97 percent of the floodwater is lost through evaporation and transpiration. A small proportion spills over into the Boteti River and flows east to the Makgadikgadi pans (the largest salt pans in the world) or southwest to Lake Ngami.

The vegetation of the delta is rich and varied, ranging from water lilies and papyrus to floodplain forests. These habitats are home to a wealth of bird life, including little bee-eaters (*Merops pusillus*), jacanas (*Actophilornis africana*), malachite kingfishers (*Alcedo cristata*), gray herons, egrets and the African fish eagle (*Haliaeetus vocifer*), whose piercing cry is one of the most beautiful sounds of Africa's wetlands. The birds share the delta with over 15 species of antelope, including the shy sitatunga and the large herds of lechwe which splash across the floodplain. Hippopotamus, zebra and buffalo are all numerous, while Botswana's elephant population numbers over 60,000, and many of these frequent the delta.

Tourists and the Okavango

People are naturally an integral part of the Okavango, and have lived in and around the delta for thousands of years. As well as introducing cattle, goats, sheep, horses and donkeys to the area, the people of the Okavango have also established an efficient agricultural system, growing corn, millet, sorghum and vegetables.

Today, a large proportion of the wetland is set aside as a wildlife preserve, Moremi Game Reserve, and much of the central and upstream end of the delta is home to important wildlife populations. This is the area favored by the burgeoning tourist industry which brings visitors from the rest of the world. Many tourist operators have permanent or temporary "camps" in the wild parts of the delta. Photographic, wilderness tourism is combined with licensed safari hunting to form the main industry inside the delta. These activities are designed under strict control so ensuring that tourism remains a sustainable industry that does not destroy its resource base.

A subsistence activity that has grown into one of commercial importance is the gathering of wetland products for village use and for sale in Maun, the largest town at the southern end of the delta. Water lily tubers, cattail roots and palm hearts are collected for food, and palm wine is made from the sap of *Hyphaene* palms. But in recent years a lucrative small industry has grown up making baskets woven from the young leaves of *Hyphaene* palms. Similarly reeds, sedges and grasses are harvested to supply fence, wall and roofing materials. Long stems of *Phragmites* reeds are gathered in the delta and along the Boteti River, and then packed into bundles before being brought to town for sale. As long as the harvesting remains sustainable, the local population will be able to make a reasonable income indefinitely.

Over the course of the past two decades the Okavango has faced many proposals to use its water for industrial, urban and agricultural purposes. In the face of these threats, all of which would have impacted the flooding regime and biodiversity of the delta, national and international concern has led to intensive scrutiny of all proposed developments. Until now these have resulted in the conclusion that the best option is to leave the water in the delta so that the wider range of benefits that it provides to biodiversity and the people who use it can be maintained. However plans for damming and water abstraction on the Okavango river in Namibia continue to be a major source of concern for the delta.

▼ *African buffalo* (Syncerus caffer).

Northern Asia

Legend:
- Ramsar Sites
- Ramsar Sites/ Parks and Reserves
- Water Bodies
- Freshwater Marsh/ Floodplains
- Tidal/Coastal Wetlands
- Deltas
- Peatlands
- Salt Pans
- Deltas

Map Note

The peatland areas of northern Asia have not been mapped as sufficiently accurate maps are unavailable.

ARCTIC

0 km   500
0 miles   250

N

Kara Sea

Barents Sea

Tanamunningen

Pasvik Nature Reserve

Kandalaksha Bay

White Sea

Islands in Onega Bay

Islands in Ob Estuary, Kara Sea

Lower Dvuobje

Patvinsuo National Park

FINLAND

Lake Ladoga   Lake Onega

Svir Delta

Kurgalsky Peninsula

Sankt-Peterburg

Mshinskaye wetland system

Pskovsko-Chudskaya Lowland

Upper Dvuobje

URAL MOUNTAINS

Ob

Ob

Irtysh

Tobol

R U S S I A N   F E D E R

Kama-Bakaldino Mires

Moscow

Oka and Pra River Floodplains

Minsk

BELARUS

Desna River Floodplains

Kiev

UKRAINE

Volga

Dnipro-Oril Floodplains

MOLDOVA

2   8   17 18 19 21
1   6 7   11 13
3   9 10 12   20
4   16
5   14 15

Black Sea

Veselovskoye Reservoir

Astrakhan

Kuban Delta

Lake Manych-Gudilo

Volga Delta

Caspian Sea

Tobol-Ishim Forest-steppe

Omsk

Chany Lakes

Lower Bagan

Astana

Kourgaldzhin and Tengiz Lakes

Lakes of the lower Turgay and Irgiz

K A Z A K H S T A N

Aral Sea

216

OCEAN

Gorbita Delta ■

◆ Area between the Pura and Mokoritto Rivers

◆ Brekhovsky Islands in the Yenisei estuary

Yenisey

ATION

Lake Uvs ■
Lake Achit ■
Ayrag Nuur ■
Har Us Nuur ◆
National Park

MONGOLIA

CHINA

Stretching over 4,000 miles (6,000 kilometers) from the Ural Mountains in the west to the Pacific coast in the east, and 3,000 miles (5,000 kilometers) from the Arctic Ocean to the mountains of Afghanistan, northern Asia includes some of the most extensive wetlands in the world. Mirroring the higher latitudes of North America, the tundra of northern Asia presents a landscape of fens and bogs interspersed with lakes, river valleys and coastal estuaries. To the south, peat bogs are the main wetland feature of the predominantly mountainous forest zone. However, extensive productive wetlands exist along river floodplains. In western Siberia, for example, the floodplain of the River Ob extends over 20,000 square miles (50,000 square kilometers). It supports the largest waterfowl breeding and molting area in Euroasia and is home to the Siberian crane (*Grus leucogeranus*) and white-tailed eagle (*Haliaeetus albicillla*).

Further south, the steppe and desert regions surprisingly support many freshwater lakes with extensive reed stands and a variety of other aquatic plants. These regions also comprise several salt lakes. The largest number of wetlands lies in the south of the Siberian lowland and in northern and central Kazakhstan. Here, flamingo, white-headed duck (*Oxyura leucocephala*), mute swan (*Cygnus olor*), ruddy duck (*Oxyura jamaicensis*), shelduck (*Tadorna tadorna*), great white and Dalmatian pelicans (*Pelecanus onocrotalus* and *P. crispus*) all nest in the wetlands.

In the high mountains of north central Asia, wetland ecosystems are largely confined to oligotrophic lakes (rich in oxygen, poor in nutrients) and fast mountain streams. The birdlife is, however, unusual, with ibisbill (*Ibidorhyncha struthersii*), bar-headed goose (*Anser indicus*) and brown-headed gull (*Larus brunnicephalus*) being typical inhabitants of the region.

At the end of the breeding season there are estimated

Key to numbered Ramsar sites
MOLDOVA
1   Lower Dniester (Nistru de Jos)

UKRAINE
2   Tyligulskyi Liman
3   Shagany-Alibei-Burnas Lakes System
4   Sharsk Lakes
5   Kyliiske Mouth
6   Dniester-Turunchuk Crossrivers Area
7   Tendrivska Bay
8   Dnipro River Delta
9   Yagorlytska Bay
10 Karkinitska and Dzharylgatska Bays
11 Big Chapelsk Depression
12 Central Syvash
13 Eastern Syvash
14 Karadag
15 Cape Opuk
16 Cape Kazantyp
17 Molochnyi Liman
18 Obytochna Spit and Bay
19 Berda River Mouth and Berianka Spit and Bay
20 Bilosaraiska Bay and Spit
21 Kryva Bay and Spit

to be over 40 million waterfowl using the wetlands of northern Asia. The tundras of the northeast are especially important and provide nesting habitats for the emperor goose (*Philacte canagica*), common, spectacled and Steller's eiders (*Somateria mollissima, S. fischeri* and *Polysticta stelleri*) and snow geese (*Anser caerulescens*), while the Taymyr Peninsula provides the breeding grounds for geese such as brent, red-breasted (*Rufibrenta ruficollis*), white-fronted (*Anser albifrons*) and bean geese (*A. fabalis*). Bird diversity is highest in the eastern regions, where European and northern Asian species mix with American species from the northeast, and Chinese and Indo-Malaysian species from the southeast. Baikal and falcated teal (*Anas formosa* and *A. falcata*) and velvet scoter (*Melanitta nigra*) are widespread breeding species, while in the Amur, swan goose, Baer's pochard (*Aythya baeri*), eastern white stork (*Ciconia boyciana*), Manchurian cranes (*Grus japonensis*) and hooded cranes (*G. monacha*) form a rich breeding avifauna.

Key to numbered Ramsar sites

MONGOLIA
1 Terhiyn Tsagaan Nuur
2 Ogii Nuur
3 Valley of Lakes
4 Khurkh-Khuiten
5 Lake Ganga
6 Lake Buir
7 Mongol Daguur

CHINA
8 Dalai Lake National Nature Reserve
9 Xianghai
10 Zhalong
11 Xingkai Lake National Nature Reserve
12 Honghe National Nature Reserve
13 San Jiang National Nature Reserve

RUSSIAN FEDERATION
14 Selenga Delta
15 Torey Lakes
16 Zeya-Bureya Plains
17 Khingano-Arkharinskaya Lowland
18 Lake Bolon and the mouths of the Selgon and Simmi Rivers
19 Lake Udyl and the mouths of the Bichi, Bitki and Pilda Rivers
20 Lake Khanka
21 Moroshechnaya River
22 Utkholok
23 Karaginski Island
24 Parapolsky Dol

JAPAN
25 Kutcharo-ko
26 Miyajima-numa
27 Utonai-ko
28 Kushiro-shitsugen
29 Kiritappu-shitsugen
30 Akkeshi-ko and Bekambeushi-shitsugen

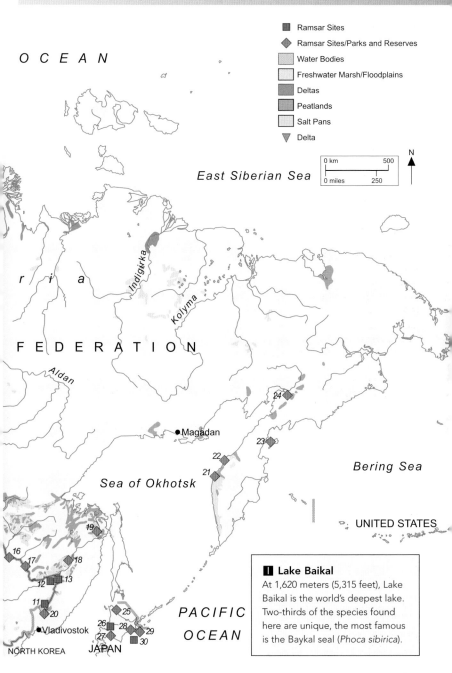

Ramsar Sites

Ramsar Sites/Parks and Reserves

Water Bodies

Freshwater Marsh/Floodplains

Deltas

Peatlands

Salt Pans

Delta

O C E A N

East Siberian Sea

0 km   500
0 miles   250

N

r   i   a

Indigirka

Kolyma

F E D E R A T I O N

Aldan

24

•Magadan

23

22

21

Bering Sea

Sea of Okhotsk

UNITED STATES

19

16

17   18

12   13

11

20

25

PACIFIC

26
28   29
30

•Vladivostok   27

OCEAN

NORTH KOREA   JAPAN

▮ Lake Baikal
At 1,620 meters (5,315 feet), Lake
Baikal is the world's deepest lake.
Two-thirds of the species found
here are unique, the most famous
is the Baykal seal (*Phoca sibirica*).

Today's threats

For centuries, the wetlands of the region avoided any serious human impacts. However, today, oil and gas exploration in western Siberia have resulted in significant pollution and transformation of the landscape. In the east, forestry and mining activities have had significant impacts upon the flora and fauna of the river basins.

In the steppe region, intensive grazing and expanding agricultural activities in the catchment, diversion of water for irrigation and pollution by fertilizers and pesticides, as well as excessive fishing in many of the lakes, are all widespread problems which need to be addressed. The degradation of the wetlands has been compounded by the drier conditions since the 1850s. In the past, the presence of very wet years with high-water levels was a critical element in preventing siltation and in maintaining the productivity of the lakes. Today, with lower flood levels this process of regeneration has been halted and the productivity of the wetlands is declining.

The Caspian region

The shallow waters of the shoreline of the Caspian Sea support an abundant waterfowl population and a rich fish fauna all year round. The highest natural production of sturgeon, for example, is from the Caspian, which

▼ *Russian Orthodox nuns check a net before fishing on the Irtysh river near Pavlodar in northern Kazakhstan.*

holds 25 percent of the world's sturgeon species. The Caspian Sea's most important wetland system in the area is the Volga Delta, which in this generally arid region supports more than 250 bird species, 60 species of mammal, including the endemic Caspian seal (*Phoca caspica*), and 80 species of plants. West of the delta, the shallow sea supports extensive underwater meadows of the grass *Chara nitella*, which is the main food plant of the 350,000 or so mute swans that visit the area in summer and fall. A rich diversity of fish-eating birds depends upon the productive fish resources of the Caspian shores. Approximately 40,000 pairs of great black-headed gulls (*Larus icthyaetus*), 80 percent of the world's entire population, breed in the Caspian wetlands, together with approximately 6,000 pairs of sandwich terns (*Sterna sandvicensis*) and 250 pairs of Dalmatian pelican.

The Bi-Ob' area
The lower Ob' valley forms a labyrinth of intricately arranged channels and floodplain lakes, intermingled with meadows and shrubs. The valley stretches 445 miles (740 kilometers) between its junction with the Irtysh River and its estuary at the head of a long fjord stretching to the Kara Sea. Referred to as the Bi-Ob' region, this vast floodplain was periodically inundated by the sea as recently as 20,000 years ago. As the sea gradually receded, the Bi-Ob' developed as a long, stretching pseudodelta, whereby a deltalike feature was formed by the sea receding rather than the river depositing silt. Today, this delta is the largest single breeding area for waterfowl in Eurasia. At the end of the breeding and molting season some 6 million birds use the area, while several hundred thousand shorebirds, ducks and geese stop here before they continue migrating.

In common with other seasonal floodplains, the Ob' region is a land of fluctuating waterlevels, with both seasonal and annual fluctuations in river discharge and flooding patterns. These flood cycles have a pronounced effect upon the wildlife of the area. In years of moderate flooding, breeding and molting waterfowl are widely distributed within the Bi-Ob' area, while in years of low floodings they move after nesting to molt in the Ob' estuary. At times of exceptionally high water levels, breeding is confined to higher, peaty areas in the floodplain and to small tributary valleys, while nonbreeding birds usually move south into the forested steppe.

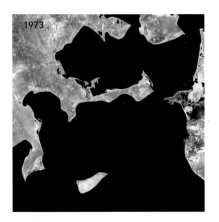
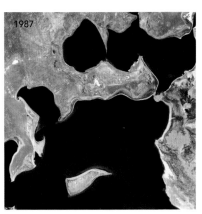

Oil exploitation

Like its United States counterpart on the Alaskan North Slope, the Bi-Ob' system is situated at the center of an oil- and gas-rich province. Great care will be needed if the exploitation of these resources does not lead to pollution of the river system and to a loss of diversity and reduced productivity of the fish and waterfowl populations of the region.

Aral Sea

Located in the heart of arid Central Asia, the Aral Sea was until recently the fourth largest of the world's lakes. Its shores were broken by the extensive deltas of the Syr and Amu rivers, two of the most biologically diverse wetland systems in the region. Today the Aral is much changed. Its surface area has been reduced from 25,000 square miles (64,500 square kilometers) before the 1950s, to around 8,500 square miles (22,000 square kilometers).

The degradation of the Aral Sea is regarded as one of the world's worst ecological catastrophes. For thousands of years the people of Central Asia used the waters of the Syr and Amu rivers to grow crops in desert oases, to which they diverted water through irrigation. These relatively small interventions had little impact on the flow into the Aral. This changed, however, in the 1960s, with the construction of much larger canals designed to irrigate land hundreds of kilometers away. As a result, the areas of irrigated agriculture in Central Asia and Kazakhstan grew from 110,000 square miles (290,000 square kilometers) in 1950 to 280,000 square miles (720,000 square kilometers) by 1992; while the average

▲ *The shrinking Aral Sea. The Aral Sea is actually not a sea at all. It is an immense lake, a body of freshwater, although that particular description of its contents might now be more a figure of speech than practical fact. In the last 30 years, more than 60 percent of the lake has disappeared. The sequence of images above, acquired by Landsat satellites, shows the dramatic changes to the Aral Sea between 1973 and 2000.*

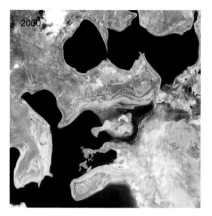

annual inflow to the sea dropped from over 11 cubic miles (50 cubic kilometers) between 1911 and 1960, to only 1 cubic mile (5 cubic kilometers) from 1981 to 1985. In 1986, a relatively dry year, no water reached the Aral Sea at all. Between 1960 and 1989, the level of the sea fell by 46 feet (14 meters), its area by 45 percent , its volume by 68 percent and the salinity of its waters increased from 1.6 to 4.5 ounces per gallon (10 to 28 grams per liter). By 1990 the Aral had divided into two parts.

Far-reaching impacts

The biological, social, and economic impact of these changes to the Aral Sea region have been disastrous. The once valuable annual fish catch has disappeared altogether, while the harvest of over a million muskrat pelts has also fallen to virtually zero. In addition, the effects of the shrinkage to the Aral Sea are not just localized. Large water bodies have a moderating effect on the neighboring climate, and today, with the decline of the sea, the climate of the region is becoming increasingly severe. It has been estimated that the growing season has declined by 10 days in the northern reaches of the Amu Darya Basin, some 120 miles (200 kilometers) away.

In the face of this array of problems there is widespread national and international concern for the future of the Aral Sea and the people of the region. In October 1990, an international symposium was held in Nukus in the Karakalpak Autonomous Soviet Socialist Republic to discuss the causes of the crisis, its impacts, and possible solutions. In their concluding resolutions, scientists argued that ecological restoration of the sea is impossible unless the hydrological cycle of the area is stabilized and reverted to something like its natural state. This could only be achieved by severely restricting the amount of water diverted from the rivers for irrigation. However, these measures will need to be accompanied by major institutional reforms which provide farmers of the region with more latitude in selecting the crops they grow and the farming practices they employ, as well as to provide greater incentives for farmers to conserve water and to reduce chemical inputs in crop production.

Central and South Asia

■ Ramsar Sites

The Himalayan mountain range dominates southern Asia. Its snowfields and glaciers give rise to some of the world's mightiest rivers, among them the Ganges, Brahmaputra, Indus, Mekong and Yangtze. For centuries, these massive freshwater arteries have nourished some of the most densely populated areas on Earth. The silt enriched valleys and huge deltas of the Indus and Ganges/Brahmaputra, for example, contain extensive wetlands that have sustained sophisticated civilizations for around 4,500 years.

Today, although modern management, technologies and practices have altered river flow and destroyed some of the different types of wetland, they continue to play a critical role in supporting the people of the continent. In Bangladesh, for example, more than 5 million people are dependent on fishing for their livelihood. The annual harvest of fish, crustaceans and frogs is estimated as being between 675,000 and 725,000 tons, of which 81 percent comes from rivers and wetlands, while the remainder is derived from marine fisheries. In India, the fishery of Chilka Lake in Orissa State alone yields 700 tons of fish per year, and provides the principal livelihood for people living along its shores.

High in the Himalayas, the main wetlands are the mountain lakes, some of which have fringing marshes. Although most of these lakes are poor in nutrients and support only a limited flora and fauna, they are of great cultural importance and are held sacred by the Hindu and Buddhist religions.

In Afghanistan, the combination of mountainous relief over much of the country and the arid climate of the southwest have confined wetlands to the major rivers which rise in the mountain ranges of the north. The largest single wetland is the Hamoun-e Puzak, one of a group of three large freshwater lakes in the Seistan Basin – an inland drainage basin surrounded by desert and lying on the border with Iran. These wetlands receive most of their water from the Helmand River, which rises far away to the northeast in the Hindu Kush. The only other large wetlands are two salt lakes in the eastern highlands, Dashte Nawar and Ab-i Estada, both of which are renowned for their flamingos.

By 1979, four reserves with important wetlands – Ab-i Estada Waterfowl Sanctuary, Dashte Nawar Waterfowl Sanctuary, Bande Amir National Park and Kole Hashmat Khan Waterfowl Sanctuary – had been established. However, conservation activities were brought to an abrupt

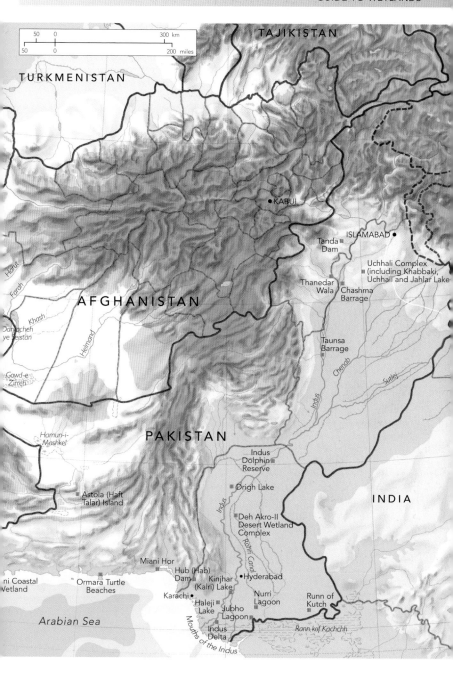

50 0   300 km
50 0   200 miles

TAJIKISTAN

TURKMENISTAN

•KABUL

ISLAMABAD •

Tanda
Dam

Uchhali Complex
(including Khabbaki,
Uchhali and Jahlar Lake)

Thanedar
Wala   Chashma
       Barrage

AFGHANISTAN

Taunsa
Barrage

Harut

Farah

Khash

Daryacheh
ye Seistan

Gowd-e
Zirreh

Helmand

Chenab

Sutlej

Indus

Hamun-i-
Mashkel

PAKISTAN

Indus
Dolphin
Reserve

Astola (Haft
Talar) Island

Drigh Lake

INDIA

Indus

Deh Akro-II
Desert Wetland
Complex

Rohri Canal

Miani Hor

ni Coastal
Vetland

Ormara Turtle
Beaches

Hub (Hab)
Dam   Kinjhar
      (Kalri) Lake   •Hyderabad

Karachi•

Haleji
Lake   Jubho
       Lagoon

Nurri
Lagoon

Runn of
Kutch

Arabian Sea

Indus
Delta

Mouths
of the Indus

Rann kof Kachchh

225

halt in 1979 due to political unrest, and it is doubtful if any practical conservation measures have been implemented since then.

Further south, in Pakistan, the main wetlands lie along the Indus Valley. These range from fresh to slightly brackish lakes and ponds (many of which provide water for urban consumption), saline and freshwater marshes, which have been formed by water from irrigation canals, to the estuary and delta at the mouth of the Indus. These wetlands are of critical importance across the country, but the pressure for land and water along the Indus is placing them under increasing pressure.

■ Ramsar Sites

■ The Mahaweli Ganga

The most extensive floodplain system in Sri Lanka is the marsh area of the Mahaweli Ganga, the largest river in Sri Lanka. The region comprises small, individual wetlands known as "villus". Twenty-six of these villus, the largest of which is 8 sq km (3 sq miles), lie within the boundaries of national parks, and constitute a linked system of protected areas. The floodplains provide a migratory corridor between wet and dry season feeding grounds for the largest concentration of elephants (*Elephus maximus*) and they are home also to the highest density of large mammals in the country, including the endan-

gered leopard (*Panthera pardus*), toque macaque (*Macaca sinica*) and sloth bear (*Melursus ursinus*). Reptiles include the threatened python (*Python molurus*), estuarine crocodile (*Crocodylus porosus*), marsh crocodile (*C. palustris*) and the endemic lizards (*Calotes zeylonensis* and *Otocryptis weigamanni*), while amphibians include the palm-frond frog (*Hylerana gracilis*). Most of the 250 resident species of birds have been recorded from this floodplain system, as well as 75 species of migratory birds. Dams constructed on the Mahaweli have altered the flow to several of the villus, but despite initial concerns they have retained their biological importance.

◀▲ Leopard

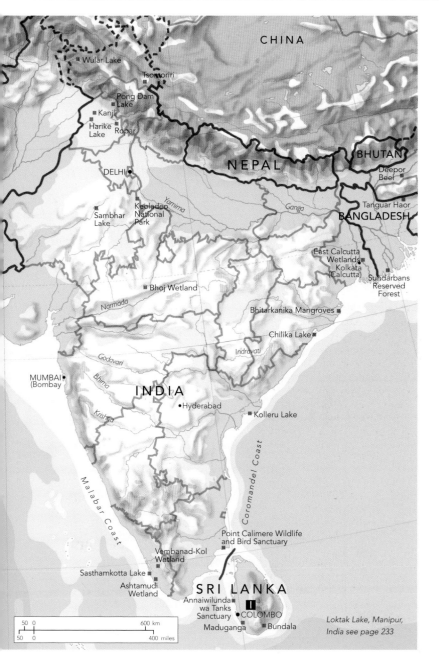

CHINA

Wular Lake

Tsomoriri

Pong Dam Lake

Kanjli

Harike Lake   Ropar

NEPAL

BHUTAN

Deepor Beel

DELHI

Yamuna

Ganga

Tanguar Haor

BANGLADESH

Sambhar Lake   Keoladeo National Park

East Calcutta Wetlands
Kolkata (Calcutta)

Sundarbans Reserved Forest

Bhoj Wetland

Narmada

Bhitarkanika Mangroves

Chilika Lake

Indravati

Godavari

MUMBAI (Bombay)

Bhima

INDIA

Krishna

•Hyderabad

Kolleru Lake

Coromandel Coast

Malabar Coast

Point Calimere Wildlife and Bird Sanctuary

Vembanad-Kol Wetland

Sasthamkotta Lake

Ashtamudi Wetland

SRI LANKA

Annaiwilundawa Tanks Sanctuary   •COLOMBO

Maduganga   ■Bundala

Loktak Lake, Manipur, India see page 233

50 0
50 0   600 km
         400 miles

India's wetlands

India is the giant of the subcontinent and with a population of over 800 million people is the world's second most populous nation. Most of the people are farmers and are concentrated in the fertile plains and valleys which support the country's principal wetlands. These include the reservoirs of the Deccan Plateau in the south, together with the lagoons and other wetlands of the southern west coast; the vast saline expanses of Rajasthan, Gujarat and the Gulf of Kutch; freshwater lakes and reservoirs from Gujarat eastward through Rajasthan and Madhya Pradesh; the delta wetlands and lagoons of India's east coast; the freshwater marshes of the Gangetic Plain; the floodplain of the Brahmaputra and the marshes and swamps in the hills of northeast India and the Himalayan foothills; the lakes and rivers of the montane region of Kashmir and Ladakh; and the mangroves and other wetlands of the island arcs of the Andamans and Nicobars, which lie about 800 miles (1,300 kilometers) off the east coast of India.

▼ *Fishing nets used for traditional fishing in the shallow waters of the bay, Chilika Lake, India.*

▲ *Wetlands cover 15 percent of Sri Lanka's land surface, ranging from man-made irrigation tanks and natural freshwater marshes inland, to lagoons and mangroves on the coast.*

Sri Lanka's wetlands extend over 25,000 square miles (65,000 square kilometers), 15 percent of the land surface of the island. The wetlands comprise both natural and man-made systems. The natural wetlands include floodplains, marshes, estuaries, lagoons and tidal mud flats, while irrigation reservoirs, rice paddies and their associated canal networks cover twice the area of the natural lakes. This wetland landscape is dominated by the 103 rivers that radiate from the wet zone, and the estuaries and lagoons which they form on the coast. These rivers support a series of natural floodplains and in the dry zone supply irrigation reservoirs through a network of canals.

Diverting the Indus

For the last 6,000 years, from the Indus civilizations of Moenjodaro to the present day, the River Indus has been the lifeblood of the arid region that is now Pakistan. However, while earlier peoples used the river's waters to cultivate the natural floodplains of the Indus, the past 100 years has seen the Indus progressively dammed and its waters diverted into one of the largest and most complex irrigation systems in the world. By 1992, there were three storage reservoirs, 16 dams, more than 40 canals totaling 35,000 miles (56,000 kilometers), and more than 900,000 miles (1.5 million kilometers) of farm channels and watercourses.

With these investments the Indus has become the "bread-basket" for most of modern Pakistan's 100 million or so people. However, the future of this productivity is threatened. In the absence of a drainage system to remove irrigation water, exacerbated by leakage of water from the banks of unlined canals, the water table has been rising in many parts of the valley. When the water table reaches the surface, it evaporates, leaving the salts carried with it in the topsoil. This salinization of the soil has, together with waterlogging, resulted in the loss of 150 square miles (400 square kilometers) of irrigated land each year and a total of 22,000 square miles (57,000 square kilometers) are now affected by salinity.

Land loss

The diversion of water is also threatening the future of the 2,300 square miles (6,000 square kilometers) of the Indus Delta. Presently 72 percent of the Indus' water is withdrawn for irrigation, leaving only 28 percent to be discharged below the Kotri, the lowest barrage on the river. Because most of this flow to the delta occurs during the monsoon (June to September), the Indus does not flow out into the sea for the rest of the year. The dams also retain the silt carried by the river and only 100 million tons (25 percent) now reaches the delta. As a result, the front edge of the delta is beginning to erode, and future sea-level rise will only exacerbate the gradual eating away of the delta.

Combined with high evaporation, the reduced fresh-water flow has raised salinity in many of the delta's creeks to 40–45 parts/1,000, which is higher than seawater (35 parts/1,000), while soil salinity is as high as 70 parts/1,000. As a result, the future of the delta's 1,000 square miles (2,600 square kilometers) of mangrove forest looks uncertain. The high salinity stunts growth and kills seedlings. In turn the biological productivity of the delta, and in particular the fish and crustaceans which use the mangroves as a nursery area, are in danger of dying out.

The Sundarbans

In a land where three of Asia's mightiest rivers, the Ganges, Brahmaputra and Meghna, mingle before flowing into the Bay of Bengal, the Sundarbans constitute the single most extensive mangrove forest in the world. Straddling the border between India and Bangladesh, the forest covers around 2,300 square miles (6,000 square

▲ *Mangrove Forest, Bangladesh. Straddling the border between India and Bangladesh, the Sundarbans form the world's largest stand of mangroves. They provide a unique haven for important populations of wildlife in this densely populated region, while also sustaining the livelihoods of over 30,000 people who harvest resources from the forest.*

kilometers). The forest floor is threaded with a complex network of rivers, creeks and canals which flood twice daily as the tide rises, creating a rich habitat for the many species of fish and invertebrates that move into the forest with the tides.

A wealth of wood

Twenty-seven species of mangrove tree grow in this dynamic system, with *Heritiera fomis* and *Excoecaria agallocha* in single or mixed stands covering more than 70 percent of the forest. In a region where the remaining areas of forest are under pressure from the rapidly growing human population, these mangroves make up 45 percent of the country's productive natural forest and are by far the single largest source of both wood and other forest products. Each year, the forest yields 2.4 million cubic feet (68,000 cubic meters) of timber, 6.4 million cubic feet (183,000 cubic meters) of pulpwood and 106,400 tons of fuelwood. There is a rich fish catch all year round, averaging about 150,000 tons, and during the winter months about 10,000 fishermen gather at the island of Dubla for fishing within the forest area and adjacent

231

coastal waters. Other products include about 3,000 tons of shells which are converted into lime, and about 200 tons of honey and 50 tons of beeswax which are harvested each year. Over 300,000 people use the forest each year in pursuit of these diverse resources with many others employed in the wood processing industry.

Despite this heavy use of the forest, it continues to support a diverse fauna of 35 species of reptile, over 270 birds and 42 mammals. Most famous among the mammals is the Royal Bengal tiger (*Panthera tigris*) for which the Sundarbans is the last remaining stronghold. It is estimated that between 350 and 450 tigers are still present in the Bangladesh portion of the forest, with a further 250 to 300 in India.

Wetlands and floods in Bangladesh

About 80 percent of Bangladesh, an area of some 44,000 square miles (115,000 square kilometers), is formed by the floodplains of the Ganges, Brahmaputra and Meghna rivers. In an average year, 10,000 square miles (26,000 square kilometers) of the floodplain are submerged, while in the largest floods of recent times this rose to 32,000 square miles (82,000 square kilometers) – 57 percent of the whole country. In addition to this massive floodplain, there are about 700 rivers and lakes which are known locally as *haores* and *beels*, and 2,400 square miles (6,100 square kilometers) of estuarine wetlands, with the Sundarbans mangrove forest forming the single most important system.

Bangladesh's wetlands play a major role in the lives of the people and economy of this very densely populated country. Availability of water for irrigation in the dry season makes it possible to grow up to three crops a year in certain areas. This enables Bangladesh's relatively small agricultural land area to produce most of the basic food requirements for the country's 110 million people. In addition, the deposition of water-borne sediments keeps the soil fertile and in some cases can even enrich it. Crop production, therefore, remains

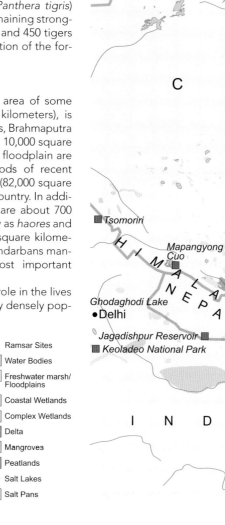

▩	Ramsar Sites
☐	Water Bodies
☐	Freshwater marsh/Floodplains
☐	Coastal Wetlands
☐	Complex Wetlands
▨	Delta
▨	Mangroves
▨	Peatlands
░	Salt Lakes
☐	Salt Pans

0 km   500
0 miles   250

N

Ürümqi

MONGOLIA

Valley of Lakes

Gobi Desert

INNER MONGOLIA

h a n

Qarqan He

H   I   N   A

GANSU

Niaodao (Bird Island)

K u n l u n   S h a n   QINGHAI

Huang He

Lanzhou

Eling Lake

Zhaling Lake

Jinsha Jiang

SICHUAN

Qinghai-
Tibet Plateau

Maidika

Nu Jiang

Chengdu

Yarlung Zangbo Jiang   Lhasa

Y A   Bitahai

L   Napahai   Dashanbao

Katmandu

Beeshazar and
Associated Lakes   BHUTAN   Lashihai   GUIZHOU

Koshi Tappu   Brahmaputra

Thimphu

Deepor Beel   Kunming

Ganga

Tanguar Haor

BANGLADESH   Loktak Lake   Y U N N A N

I   A

Dacca

Calcutta   East Calcutta Wetlands   MYANMAR   VIETNAM

Sundarbans Reserved Forest

Bay of Bengal   (BURMA)   LAOS

Bhitarkanika Mangroves

233

at a satisfactory level without the need for potentially harmful fertilizers. Flooding also promotes algal growth which transfers up to 27 pounds of nitrogen per acre (30 kilograms per hectare) per year from the atmosphere to the soil. In addition, fish provide somewhere in the region of 80 percent of the daily animal protein intake of the people, while the fishery sector's contribution to the country's Gross Domestic Product (GDP) and export earnings is 6 and 12 percent respectively. The sector also generates full-time employment for about 2 million people and part-time employment for another 10 million.

Flood control
Despite the importance of these extensive wetlands, and the close dependence upon them of the people of Bangladesh, the country is better known outside of the Asian subcontinent for the damage inflicted upon it by floods and cyclones. Since Bangladesh was established in 1971, a series of massive floods and devastating cyclones have killed more than 500,000 people and left millions homeless.

In response the international community is currently investing about US$150 million to prepare proposals for investment to control the floods and protect people from cyclones. Yet, with 55 percent of the land surface

lying less than 10 feet (3 meters) above sea level at the mouth of three of the world's largest rivers, Bangladesh is a land of floods. The people of the delta have adjusted to this age old phenomenon, and live with the rhythm of the flood. Indeed, in a country of 110 million people where fertile, agricultural land is at a premium, farmers await the flood with anticipation because it brings nutrients and water for agriculture. However, the close relationship between the farmers and the flood is becoming strained. Evidence gathered over the last 40 years suggests that extreme floods are becoming increasingly frequent. Since 1954, severe floods have submerged an area of between 13,500 square miles (35,000 square kilometers) and 32,000 square miles (82,000 square kilometers) in 13 years. In 1988, when 32,000 square miles (82,000 square kilometers), 57 percent of the country, was flooded, over 2,000 lives were lost and damage to property and infrastructure was estimated at US$1.1 billion. The dilemma facing the Bangladesh Government is how best to invest in reducing the effects of these events.

Working with water

Current proposals focus upon the construction of dikes and other structures which would close off valuable land and habitation from the flood. However, this approach has been the subject of strong criticism from a wide audience, including some highly respected international authorities on water management. Some argue that, rather than invest huge sums in controlling nature, it would be best to work with nature. They point to the fact that in 1757 the region already had a system of small dikes which prevented major damage. Then people simply took to boats if the waters got too high, while benefiting from the fertilizing role of the flood when the water receded.

The issues are certainly complex. The Government of Bangladesh faces the dilemma of needing to protect its population from flood and minimizing damage to infrastructure, while at the same time ensuring that the benefits of the flood are not altogether lost. However, the issue of debate is whether this is indeed possible in a river system, the combined flow of which is second only to the Amazon, and the position of which has been known to move several kilometers in a season. In addition, some authorities estimate that while the urban rich will benefit, over 6 million of the country's poorest people would suffer.

◀ Bangladeshis fish in floodwater at Chhatak in Sunamganj 250 miles (400 km) northeast of the Bangladeshi capital Dhaka in July 2004. In some areas of flood-hit Bangladesh, floods submerged crops and cut rail and road links, pushing up food prices and forcing thousands of people to cram into government food shelters.

East Asia

With a population of 1.3 billion projected to rise to 2 billion by the year 2025, East Asia currently accounts for approximately 24 percent of world population. For centuries, the high productivity of East Asia's wetlands, which today still cover some 108,000 square miles (280,000 square kilometers), has played a central role in sustaining the living standards of hundreds of millions of people. For example, almost 95 percent of China's population of 1.1 billion is concentrated in the eastern half of the country, principally in the vast alluvial plains of the major rivers.

China has more than 97,000 square miles (250,000 square kilometers) of wetlands. These include marshes and bogs, lakes (both natural and artificial), and coastal salt marshes and mud flats. This diversity can be grouped into five principal systems: the high-altitude lakes and bogs of the Qinghai-Tibetan Plateau, the inland drainage systems of the Sinkiang Basin, the freshwater lakes, marshes and peat bogs of Heilongjiang, Jilin, Lianing and Inner Mongolia in the northeast; the lake systems and riverine wetlands along the middle and

■ Ramsar Sites

Lake Uvs and its surrounding wetlands
Lake Achit and its surrounding wetlands
Tesyn Gol
Selenge Moron
■ Ayrag Nuur
Orhon Gol
Har Us Nuur National Park
Terhiyn Tsagaan Nuur
Lakes in the Khurkh-Khuiten Valley ■
■ Ogii Nuur
ULAN BATOR

MONGOLIA

Valley of Lakes ■
(Boon Tsagaan Nuur,
Taatsiin Tsagaan Nuur,
Adgiin Tasgaan Nuur,
Orog Nuur)

CHINA

100   0
100   0   600 km
                           400 miles

Eerduosi National
Nature Reserve
■

Mongolia's lakes

Most of Mongolia's principal lakes are situated in the Central Asian Internal Drainage Basin. Some 144 million years ago, this vast basin formed an inland sea which covered the entire region. As the humid climate of these prehistoric times gave way to more arid conditions, however, the sea gradually broke up to form the relict lakes that remain today. These include freshwater lakes with outlets into others such as Ayrag and Har-Us lakes, and saline lakes such as Hyargan and Dorgon which have no outlet. The fish fauna of the basin reflects its history. Many are relict species that are left over from the ancient sea.

lower reaches of the Yellow and Yangtze rivers and finally the estuarine, mud flat and mangrove systems along the coast.

China's 8,000 square miles (21,000 square kilometers) of coastal marshes and mud flats are concentrated in three main areas: at the mouth of the Yangtze River and along the adjacent coast of Jiangsu Province, around the estuary of the Yellow River in Bohai Gulf and in the estuarine system of the Shuangtaizi and Liao rivers in Liaoning Province. Only a small proportion of what are thought to have been extensive wetlands now remain. Scant stands of mangroves grow along the coast as far north as central Fujian Province and also on the island of Taiwan, but many of these mangrove/mud flat ecosystems have now been converted to

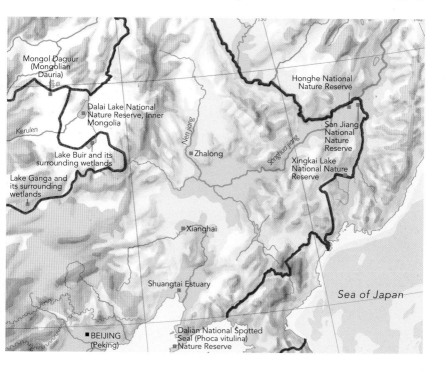

rice paddies and aquaculture ponds. One of the richest surviving stands of mangrove is in Deep Bay, on the border between Guangdong Province and Hong Kong.

Diverse wetlands

Despite rainfall that averages only 4–16 inches (100–400 millimeters) a year, Mongolia is rich in water resources mainly because of the high mountain ranges which attract precipitation. There are approximately 6,000 square miles (15,000 square kilometers) of standing water bodies and 30,000 miles (50,000 kilometers) of rivers. Wetland habitats are extremely diverse, ranging from cold, deep ultraoligotrophic (poor in nutrients and rich in oxygen) lakes to temporary saline lakes, and there are many major rivers possessing extensive floodplains. Only the southern desert margin and the southeast of the country lack any permanent water. Many of the lakes and marshes are extremely important breeding and staging areas for migratory waterfowl, notably ducks, geese and cranes. Some of these birds enjoy special protection that dates back to the 1200s when laws were enacted forbidding the hunting of game during the breeding season. In

▼ *Lake Biwa, Japan. Lake Biwa is the largest freshwater lake in Japan and has played an important role in the country's history. For many years it was a major transportation artery between Japan's west coast and the cities of Kyoto and Osaka to the east. Today the lake is a major source of water for domestic supply, irrigation, and industry. Over 50,000 waterbirds use the lake during the winter months.*

Ramsar Sites

Ramsar Sites/
Parks and Reserves

Water Bodies

Freshwater Marsh

Tidal/Coastal Wetland

Delta

Peatlands

Complex Wetlands

addition, human population pressure remains low throughout much of Mongolia and most wetlands are still in an almost pristine condition, disturbed only by the occasional hunter, fisherman or shepherd.

The wetlands of Korea and Japan

In the Korean Peninsual the principal wetlands are along the west and south coasts. Here there are numerous estuaries and shallow sea bays with extensive intertidal mud flats and offshore islands. In some areas there is a broad coastal plain with many small lakes, expansive reedbeds and large areas of rice paddy. This marshy coastal plain is especially well developed around the estuaries of the Chongchon-Gang and Teadong rivers in the north, and in the lower basins of the Imjin, Han and Nakdong rivers in the south. These wetland systems are now under heavy pressure. Of the 487 natural lakes and ponds in Japan, most are very small. However, the largest lake in Japan, Lake Biwa in central Honshu, is the notable exception. This particular lake covers 260 square miles (674 square kilometers). Eastern Hokkaido supports the most extensive freshwater marshes remaining in Japan, as well as the majority of the country's remaining lagoons and salt marshes.

Elsewhere, most of the lowland marshy habitats and coastal lagoons have been drained for agriculture. The tidal ranges of Japan's Pacific coast sustain the country's largest area of intertidal mud flats. Estuaries and bays such as Tokyo Bay and Inner Ise Bay also once supported extensive mud flats, but much of this habitat has been lost to urban development. Mangrove swamps are confined to the Amami Islands and Ryukyu Islands, where they grow on muddy beaches and in estuaries.

Lakes and the Three Gorges Dam

The greatest concentration of freshwater lakes in China occurs on the alluvial plains along the middle and lower reaches of the Yellow and Yangtze rivers. The total area of lakes is estimated at over 8,500 square miles (22,000

▼ *Water gushes from more than 20 open sluice gates on the Three Gorges dam in Yichang, Hubei Province, September 9, 2004.*

square kilometers). The most important include the Dongting Lakes in Hunan Province, the Wuhan Lakes in Hubei Province, Poyang Lake in Jiangxi Province and Shegjin Lake in Anhui Province, all in the Yangtze Basin. The lakes are used intensively by local people for fishing, grazing and farming, practices that do not threaten the internationally important populations of waterbirds, such as the Siberian crane (*Grus leucogeranus*), which utilize the lake in winter. The world population of the Siberian crane was thought to number only a few hundred until a wintering flock was discovered at Poyang Lake in 1980. Numbers are now estimated at 2,600 individuals, approximately 95 percent of the known world population of this species.

The principal threat to these lakes is the alteration of the hydrological regime, either through retention of water in dams upstream or through modification of the flow patterns. One such project is the enormous Three Gorges Dam planned for the Yangtze River. Estimated in early 1989 to cost US$10 billion, the project is expected to generate almost a fifth of China's entire energy needs. Its supporters claim that the dam will control the massive flood that devastates the floodplain of the Yangtze about once every 10 years, while also improving navigation along the Yangtze, China's most important waterway. Environmentalists, however, warn that the project will lead to flooding on a huge scale of the area behind the dam, which in turn will lead to the displacement of more than a million people. The dam is also likely to reduce the flooding of the natural wetlands along the Yangtze Valley, including Poyang Lake.

East Asia's wildlife

Extending almost from the subarctic to the tropics, the wetlands of China, Mongolia, Korea and Japan support an especially rich and diverse fauna and flora. Over 170 species of waterbird occur in the region, including 36 species of duck and goose, eight species of crane and 53 species of shorebird. Wetlands of Mongolia, the Qinghai-Tibetan Plateau, northeastern China and Hokkaido support large breeding populations of waterbirds, while the lakes, marshes and coastal wetlands of central and southern China, Korea and southern Japan provide important staging and wintering areas for these and many other migratory species of waterbird breeding in arctic Russia.

However, populations of most species of waterbirds

and other wildlife have been reduced to a fraction of their former levels. The reductions are a result of the massive loss of wetlands to agriculture and urban development, very heavy hunting pressure and general disturbance from the huge human population. No less than 27 species of waterbird are now listed in the IUCN Red Data Book as threatened, and several of these are nearly extinct.

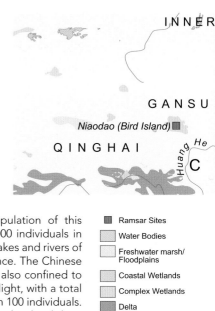

Ramsar Sites

Water Bodies

Freshwater marsh/Floodplains

Coastal Wetlands

Complex Wetlands

Delta

Mangroves

Peatlands

Salt Lakes

Salt Pans

Asia's crocodile

East Asia's only crocodilian, the Chinese crocodile (*Alligator sinensis*), is on the brink of extinction. The total population of this species was thought not to exceed 2,000 individuals in 1984, their last stronghold being in the lakes and rivers of the lower Yangtze Valley in Anhui Province. The Chinese river dolphin or Baiji (*Lipotes vexillifer*), also confined to the Yangtze River, is in an even worse plight, with a total population possibly numbering less than 100 individuals. Both species have been reduced to these low levels by a combination of direct persecution, and dam and barrage projects on the Yangtze and its tributaries. The Chinese water deer (*Hydropotes inermis*), however, remains fairly common around lakes in the Yangtze Valley and in the extensive coastal marshes of Jiangsu Province.

China's reeds

The wetlands of temperate northeastern China include some of the most extensive reed marshes in Asia. The Sanjiang (Three Rivers) Plain near the confluence of the Heilong (Amur), Sungari and Wusuli (Ussuri) rivers, is the largest single wetland area, with over 4,000 square miles (10,000 square kilometers) of shallow lakes, reed beds and peat bogs. But there are several other very extensive systems of freshwater lakes and marshes such as Zhalong Marshes near Qiqihar and Xiang Hai Marshes in western Jilin. These marshes are critically important habitats for both migratory and breeding cranes, and other waterbirds.

In several areas, reeds are harvested on a commercial basis for the production of high-quality paper. At Zhalong Marshes in Heilongjiang Province, one state-run company employing 450 full-time workers and 1,000 temporary workers harvested 26,000 tons of reeds in 1984. Most

MONGOLIA

Eerduosi National Nature Reserve

Beijing ●

Dalian National Spotted
Seal Nature Reserve

Bohai Gulf

HEBEI

SHANXI

Huang He

Huang He
(Yellow River)

SHANDONG

● Lanzhou

H I N A

Yellow Sea

HENAN

JIANGSU

Yancheng National Nature Reserve

Dafeng National Nature Reserve

ANHUI

Chang Jiang

Chongming Dongtan Nature Reserve

● Shanghai

Chang Jiang
(Yangtze River)

HUBEI

● Chengdu

Wuhan Lakes

Chongqing ●

ZHEJIANG

Dongdongtinghu

Dongting
Lakes

Poyanghu

Xi Dongting Lake
(Mupinghu) Nature Reserve

Nan Dongting
Wetland and Waterfowl Reserve

SICHUAN

HUNAN

JIANGXI

East
China Sea

GUIZHOU

FUJIAN

● Kunming

GUANGDONG

T'aipei ●

YUNNAN

Xun Jiang

● Guangzhou

TAIWAN

Mai Po Marshes and Inner Deep Bay

Hong Kong ●

Huidong Harbor Sea Turtle
National Nature Reserve

VIETNAM

Shankou Mangrove Nature Reserve

Hanoi ●

Zhanjiang Mangrove National Nature Reserve

LAOS

Xuan Thuy
Natural Wetland
Reserve

Dongzhaigang

South China Sea

Hainan

0 km   500

0 miles   250

N

of these wetlands have remained largely unaltered by human activity. However, proposals to divert the rivers that flood the Sanjiang Plain may lead to the loss of this large floodplain system within the next few years.

Coastal reclamation in Korea

A severe shortage of land suitable for agriculture in the Republic of Korea is now putting a tremendous strain on the country's coastal wetland regions. Much of the traditional agricultural land has been lost to the rapid development of urban and industrial areas in recent years. Domestic food production has been falling while at the same time the human population has been rising steadily for the last 40 years or so. As a consequence, there has been a steep rise in food imports.

In an effort to solve these problems, the Korean government is pursuing a major program of land reclamation at estuaries and shallow bays on the south and west coasts of the country. In a feasibility study carried out by the government in 1984, 155 estuaries and bays with a total area of some 1,600 square miles (4,180 square kilometers) were identified as being suitable for land reclamation projects. It is anticipated that all 155 sites will eventually be reclaimed, resulting in the loss of approximately 65 percent of the total coastal wetlands of the country.

▼ Horsemen in Ruoergai peatland, China. Lying at an altitude of 11,000–21,000 ft (3,400–3,900 m), the Ruoergai marshes are some of the world's highest altitude peatlands. They cover an area of 1,210,790 acres (490,000 ha) in the headwaters of the Yellow River. In addition to the globally important biodiversity that they support, the peatlands also help sustain the livelihoods of local people, including nomadic Tibetan herders.

The cranes of East Asia

Eight of the world's 15 species of crane live in East Asia: the common crane (*Grus grus*), black-necked crane (*G. nigricollis*), hooded crane (*G. monacha*), red-crowned crane (*G. japonensis*), white-naped crane (*G. vipio*), sarus crane (*G. antigone*), Siberian crane (*G. leucogeranus*) and demoiselle crane (*Anthropoides virgo*). The sarus crane, primarily a species of southern Asia, is an extremely rare visitor to wetlands in Yunnan Province in southern China, while the common and demoiselle cranes are relatively prevalent and widespread across much of Eurasia, spending winter in Africa and the Indian subcontinent. The other five species are, however, more or less confined to eastern Asia.

The black-necked crane breeds in marshes and peat bogs at elevations of over 11,000 feet (3,300 meters) above sea level on the great Qinghai-Tibetan Plateau in southwest China and in neighboring Ladakh in India. In fall, the cranes undertake relatively short migrations to spend the winter months in valleys and marshes at lower elevations, generally between 2,000 and 6,500 and 8,000 feet (2,500 meters) above sea level. Key wintering areas include the Lhasa Valley in Tibet, the Popshika, Boomthang and Tashi Yangtsi valleys in Bhutan, Lu Guhu, Bitahai and Napahai marshes in Yunnan Province, and Caohai (which literally translates as "sea of grass") in neighboring Guizhou Province. The total population of this rare crane has recently been estimated at about 3,000 individuals.

Migrating cranes

The other four species of crane are northern breeders, undertaking long migrations via a series of traditional staging areas to wintering grounds in eastern China, the Korean Peninsula and southern Japan. The Siberian crane undertakes the longest migration, traveling from breeding areas in the Arctic tundra of Yakutia to its wintering grounds in the Yangtze Valley. The great bulk of the 3,000 or so Siberian cranes spends the winter at Poyang Lake, although small flocks also settle on Shengjin Lake and the Dongting Lakes. Traditional staging areas include Zhalong and Momoge marshes in Heilongjiang Province in northeastern China. Two much smaller populations of Siberian cranes breed further west in arctic Russia and winter in northern India and Iran. Unfortunately, however, both these populations now number no more than about 15 individuals.

The sarus crane is one of the largest members of the crane family, and mature adults stand more than 5 ft (1.5 m) tall. The sarus crane is distributed over a 600-mile (1,000-km) wide band that stretches from the Hindu Kush to Vietnam. Among the human population of this region, the sarus crane has acquired a protected, almost sacred status. The reason for the veneration of the sarus crane is the close, lifelong pair-bonding that characterizes the species. This bonding is so close that adult cranes are rarely seen alone. To the human population, fond of animal archetypes, the sarus crane has become symbolic of fidelity, a virtue that is as highly regarded in the East as it is in the West.

The hooded crane breeds in the marshes of the coniferous-forested taiga zone in eastern Siberia and Amurland. While a small proportion of the population migrates through eastern China to winter at Shengjin, Poyang and the Dongting lakes in the Yangtze Valley, the bulk of the population takes a more easterly route through the Korean Peninsula to spend the winter in ricefields at West Taegu in South Korea and at Izumi (Kyushu) and Yashiro (Honshu) in southern Japan. By far the most important site for this species is the famous crane sanctuary at Izumi, where as many as 5,500 of the total population of about 6,500 spend the winter.

The white-naped crane breeds in the extensive wetlands of the temperate steppe zone. This area stretches from eastern Mongolia through Heilongjiang and Jilin Provinces of northeastern China to the Ussuri River along the Russian border. About two-thirds of the population, which numbers some 3,500 birds, take a westerly route to winter in the Yangtze Valley (Shengjin, Poyang and Dongting), while the remainder take an easterly route to winter in the Korean Peninsula (principally at Taesong'-dong and Panmunch'om) and at Izumi in Japan.

The red-crowned crane is the most southerly breeder of these four cranes, breeding in reed marshes in Hokkaido, Amurland and northeastern China as far south as the estuarine marshes of the Shuangtaizi and Liao rivers in Liaoning Province. The Hokkaido population, which numbers about 350 to 400 birds, is largely sedentary, the birds moving only a short distance from their breeding marshes to spend the winter around feeding sites near Kushiro. A population of about 400 birds breeding along the Ussuri River and in Amurland winters primarily in the coastal marshes of South Hwanghae and Kaesong in the Democratic People's Republic of Korea. Western breeding birds, numbering about 650, migrate round the shores of the Bohai Gulf to winter in the vast coastal marshes of Jiangsu Province and to the islands at the mouth of the Yangtze River.

Cranes and golf

Known as *tancho*, the red-crowned crane has become a symbol of conservation in Japan, expanding their traditional role as symbols of longevity and happiness. In Kushiro, one of the major threats to the cranes that breed and winter there comes from an unexpected quarter: golf courses. In the area surrounding Kushiro Marsh, there are seven golf courses and an additional 12 are under construction. Presently, the courses cover an area of some 19 square miles (50 square kilometers). In the interests of high-quality fairways and greens, the courses have been sprayed with pesticides, herbicides and fertilizers, many of which eventually find their way into the marsh downstream. While the precise effects of these chemicals are unknown, the toxic effects of pesticides in many other similar situations have given rise to growing concern for the cranes and other animal species that depend upon the marsh. Conservationists argue that it would be a national tragedy if the *tancho* was to be endangered because of golf.

▼ *Common cranes (grus grus), breeding (left) and nonbreeding (right).*

Southeast Asia

INDIA

Loktak Lake

CHINA

Song Hong

Guangzhou

MYANMAR
(BURMA)

Salween

Shankou Mangrove
Nature Reserve

Hanoi

Zhanjiang
Mangrove
NNR

Gulf of
Tonkin

Irrawaddy

Nong Bong Kai
Non-Hunting Area

Xuan Thuy
Natural Wetland
Reserve

Dongzhaigang

Hainan

Bay of
Bengal

Vientiane

LAOS

Rangoon

Bung Khong Long
Non-Hunting Area

THAILAND

South
China
Sea

Mekong

VIETNAM

Bangkok

Don Hoi Lot

Mekong River

Tonle Sap

CAMBODIA

Boeng Chmar

Andaman
Sea

Andaman
Islands

Phnom Penh

Koh Kapik

Ho Chi Minh City

Kaper Estuary – Laemson
Marine National Park

Kraburi Estuary

Gulf of
Thailand

Mu Koh Ang Thong
Marine National Park

INDIA

Pang Nga Bay
Marine National Park

Krabi Estuary

Nicobar
Islands

Had Chao Mai MNP –

Ta Libeng Island –
Trang River Estuaries

INDIAN OCEAN

MALAYSIA

Kuan Ki Sian of the Thale Noi
Non-Hunting Area Wetlands

Thale Luang

Princess Sirindhorn Wildlife Sanctuary
(Pru To Daeng Wildlife Sanctuary)

Covering 2.3 million square miles (6 million square kilometers) and straddling the Equator at the southeastern extremity of the Asian mainland, Southeast Asia is home to 500 million people from a wide variety of ethnic and cultural groups. For centuries, many of these cultures have depended closely upon the diverse wetlands of the region, which include intertidal flats and mangrove forests, freshwater and peat swamp forests, lakes, rivers and marshes. Almost all the wetlands in this region are naturally forested and so they combine the values of wetlands and tropical forests. They support more than 50 percent of the vertebrate animals of the region, while also providing flood control, water supply, fisheries and coastal protection.

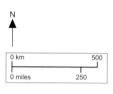

Huidong Harbor Sea Turtle NNR

Hong Kong

Mai Po Marshes and Inner Deep Bay

N

0 km	500
0 miles	250

PHILIPPINES

MALAYSIA

- ▦ Ramsar Sites
- ▢ Water Bodies
- ▢ Freshwater Marsh
- ▢ Coastal/Tidal Wetland
- ▤ Swamp Forest
- ▥ Mangroves
- ▦ Peatlands
- ▢ Complex Wetlands
- ▦ Deltas

Including the Malay Peninsula, Indochina and the 20,000 islands that form Indonesia and the Philippines, the coastline of Southeast Asia is nearly 93,000 miles (150,000 kilometers) long, four times longer than the coast of Australia. Much of this coastline is bordered by mangroves, mud flats and sandy beaches, together with coral reefs. The shallow coastal waters are extremely productive and support an abundance of life, which in turn provides food for much of the region's human population.

Swamps and swamp forest

Freshwater and peat swamp forests occur inland of mangroves in lowland areas. Freshwater swamps lie on permanently or seasonally flooded soils, normally in a zone up to 3 miles (5 kilometers) wide along rivers. This forest type is very diverse, with records of over 100 tree species recorded on a 0.25 acre (0.1 hectare) in Sumatra. Many species have buttresses, prop roots or pneumatophores to enable them to "breathe" during times of inundation. The forests are of great economic and social value to human communities in adjacent areas. In the Malaysian Peninsula alone, more than 800 plant species of value have been recorded in the freshwater swamp forests, which now cover only 20 percent of their original area. The swamp forests also play an important role in supporting freshwater fisheries as many fish species breed or feed in the forests during periods of high water. Despite their value, freshwater swamp forests have come under heavy pressure for conversion to agricultural land, particularly for ricefields and more recently for oil palm plantations.

There are relatively few lakes in the lowlands of most of Southeast Asia. It is thought that rapid siltation rates and plant growth cause lakes rapidly to become swamp forests or other wetland types. The exception to this is in the Philippines and parts of Indonesia, where volcanic activity has led to the formation of more than 100 lakes, many supporting unique species. The largest lake in Southeast Asia is Danau Toba, which covers 520 square miles (1,350 square kilometers). Danau Toba is a volcanic caldera (basin-shaped crater) lake with spectacular scenery and is an important tourist attraction. The largest lake in the Philippines is Lake Lanao in Mindanao. It covers 135 square miles (350 square kilometers) and supports 20 endemic fish species. The major threats to lakes are from forest clearance of their immediate catchment areas and from pollution by industrial or urban waste.

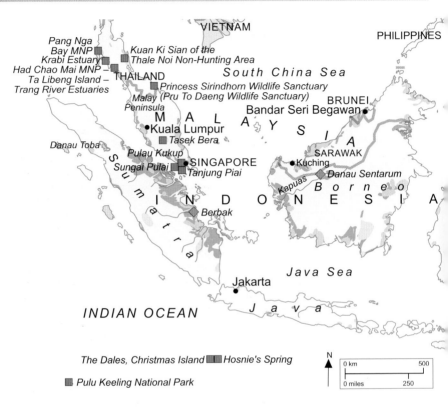

Pang Nga
Bay MNP
Krabi Estuary
Had Chao Mai MNP –
Ta Libeng Island –
Trang River Estuaries

Kuan Ki Sian of the
Thale Noi Non-Hunting Area

THAILAND

VIETNAM

PHILIPPINES

South China Sea

Princess Sirindhorn Wildlife Sanctuary
(Pru To Daeng Wildlife Sanctuary)

BRUNEI

Malay
Peninsula

M A L A Y S I A

•Kuala Lumpur

Bandar Seri Begawan•

Danau Toba

Tasek Bera

Pulau Kukup
Sungai Pulai

SINGAPORE
Tanjung Piai

SARAWAK

•Kuching

Danau Sentarum

Kapuas

B o r n e o

Sumatra

I N D O N E S I A

Berbak

Java Sea

Jakarta

INDIAN OCEAN

J a v a

The Dales, Christmas Island ▨▨ Hosnie's Spring

▨ Pulu Keeling National Park

N

0 km   500

0 miles   250

Threats and protection

Despite the wide ranges of resources and services provided through wetlands in the region, they have not generally been properly protected or managed on a sustainable basis. In Southeast Asia, major threats come from mining, aquaculture, unsustainable forestry or fishery practices, and conversion to agricultural or urban land. Threats to 94 percent of the wetlands of international importance have been reported, while moderate to high threats have been reported at over 45 percent of wetlands that lie in totally protected areas. If these systems are to continue to support the region's biodiversity and the people who depend upon their resources, urgent attention needs to be given to improving their management. Governments and nongovernmental organizations (NGOs) have been paying increasing attention to this issue in recent years, and Vietnam and Indonesia both joined the Ramsar Convention in the late 1980s. There is

Ramsar Sites

Ramsar Sites/
Parks and Reserves

Water Bodies

Freshwater Marsh

Coastal/Tidal Wetland

Swamp Forest

Mangroves

Peatlands

Complex Wetlands

▼ *Central Kalimantan, Indonesia. Many of Kalimantan's inhabitants live along the region's waterways on floating and raised houses. This allows them to adapt to the changing water levels as the rivers rise and fall in the rainy and dry season.*

a growing recognition of the value of wetlands, and a number of agencies have established wetland units or advisory groups to begin to address the problems. However, a major effort will be needed over the next few years to ensure that a proper foundation is laid for future sustainable management of the wetland resources.

The Mekong River

With a catchment area of 305,000 square miles (795,000 square kilometers) and stretching for 2,600 miles (4,200 kilometers), the Mekong is the longest river in Southeast Asia and the twelfth longest in the world. It originates in the Tibetan highlands on the southern border of Tsinghi Province, China. At the end of its long journey, the river drains into the South China Sea through its delta, which lies to the south of Ho Chi Minh City in Vietnam. The river is called Cuu Long ("Nine Dragons") by the Vietnamese after the delta's complex web of distributary channels. From just below Phnom Penh to the sea, the delta covers an area of 21,000 square miles (55,000 square kilometers), and is one of the largest and most important wetland systems in Asia.

Along its way to the sea, the Mekong is the lifeblood of the region. About three-quarters of its drainage basin lies within Laos, Thailand, Cambodia and Vietnam and about 40 percent of the total population of these four countries live in the basin. The cities of Vientiane and Phnom Penh are located on its banks, and the river plays a central role in the economy of the whole region.

Hydroelectricity

Since 1957, at least 100 hydroelectric dams, including seven major dams, have been proposed for the Mekong and its tributaries. In addition to generating electricity, the proposed dams would help to exploit the estimated 15,000 square miles (39,000 square kilometers) of potential agricultural land in the delta by providing controlled irrigation water.

So far, just over a dozen dams have been finished and several others are under construction. While these have helped increase agricultural production, negative effects are beginning to be felt. Beneficial flooding during the monsoon season has been reduced. Freshwater inflow to the coastal ecosystems, including the mangrove forests, has decreased, and salinity patterns and levels have been altered. If more of the proposed dams are constructed it is expected that the adverse effects will increase. The species of fish that use the Mekong River system for spawning grounds and migration paths will be disturbed by dam construction upstream, a continued lack of fresh-water inflow will severely degrade the mangrove forests, while a reduction in flood-borne silt will decrease soil fertility and agricultural production.

Tonle Sap Lake and Tonle Sap River

During the dry season, Cambodia's Tonle Sap (Great Lake) covers some 970–1,150 square miles (2,500–3,000 square kilometers). As the water level in the Mekong rises in June or July, the flow in the Tonle Sap River is reversed and the Mekong flood-waters enter Tonle Sap. At the height of the flood season in September and October, the lake and its inundation zone can cover as much as 5,000 square miles (13,000 square kilometers) of plains extending from the northwestern corner of the country to the Mekong at Phnom Penh. Large tracts of freshwater swamp forest grow in the floodplain and estimates in

Luzon

Manila

Naujan Lake National Park

PHILIPPINES

Olango Island
Wildlife Sanctuary

Tubbataha Reefs National Marine Park

Sulu Sea

Mindanao

Agusan Marsh
Wildlife Sanctuary

Davao

Celebes Sea

PACIFIC OCEAN

Sulwesi

Moluccas

Jayapura

IRIAN JAYA

Banda Sea

PAPU

INDONESIA

Lake Kutubu

Dili

Timor   EAST TIMOR

Arafura Sea

Tonda Wildlife
Management Area

Cabourg Peninsula

AUSTRALIA

▌ Lake Manguao

Situated in the northern part of the island of Palawan, Lake Manguao is one of very few pristine, lowland lakes left in the Philippines. Covering about 2.5 sq mi (6.5 sq km), the lake was formed by the damming of a river by lava. During the last ice age, Palawan was connected with Borneo and the fish and plant species show closer affinities with those of mainland Southeast Asia than the rest of the Philippines. The destruction of forest cover of Lake Manguao's catchment area, mainly due to the 5 percent rise in population, is threatening the region's plants and animals.

the 1970s put the total area of forest in Cambodia at about 2,600 square miles (6,800 square kilometers). But by 1990 almost 20 percent of the swamp forest had been cleared for firewood, agricultural land and fish ponds.

War and the Mekong Delta

Vietnam is facing one of its biggest challenges since the Vietnam War ended in 1975. American and Vietnamese scientists estimate that 8,500 square miles (22,000 square kilometers) of forest and a fifth of the country's farmland were affected as a direct result of bombing, mechanized land-clearing and defoliation. About 1,000 square miles (2,800 square kilometers) of the Mekong Delta was under mangrove and *Melaleuca* forest at the start of the war. From the early days of the war, the delta was extensively sprayed with defoliants, including Agent Orange, which resulted in the destruction of an estimated 480 square miles (1,240 square kilometers) of mangrove and 100 square miles (270 square kilometers) of *Melaleuca* forest. This amounted to more than 50 percent of the entire mangrove forest in Vietnam.

The exposure of the alluvial forest soil to high temperatures after trees were defoliated caused desiccation and leaching of nutrients from the soil, making it very hard and unsuitable for plants. Some of the chemicals

EW GUINEA

Solomon Sea

Port Moresby

SOLOMON ISLANDS

●Honiara

Guadalcanal

Rennell

Coral Sea

N

| 0 km | 500 |
| 0 miles | 250 |

■ Ramsar Sites

◆ Ramsar Sites/ Parks and Reserves

▢ Water Bodies

▢ Freshwater Marsh

▢ Tidal/Coastal Wetland

▢ Deltas

▢ Mangroves

▢ Peatlands

▢ Swamp Forest

▢ Complex Wetlands

remained active in the soil for many months and subsequently destroyed the soil's micro-organisms. As a result, there has been little recolonization in many areas. Efforts by the Vietnamese have, however, improved the rate of recovery; more than 270 square miles (700 square kilometers) have now been replanted.

The *Melaleuca* forests that grew on the plains behind the mangroves were burned with Napalm as well as sprayed. Furthermore, canals were dug to drain the flooded areas in another effort to flush out the guerrilla army. The dried areas became highly acidic due to chemical changes in the soil caused by oxidization and the land infertile; the *Melaleuca* forests were almost destroyed. Local and international efforts have been made to restore the forests and the fertility of the land by replanting. In the long-term, it is intended to harvest the *Melaleuca* for timber, but the short-term plan is to harvest honey from bees feeding on the rich flowers and to extract oils from the leaves.

Mangroves

Growing predominantly in Indonesia, Malaysia and Papua New Guinea, the mangroves of Southeast Asia cover more than 23,000 square miles (60,000 square kilometers) – about 1 percent of the world's total area. Containing more than 60 species of tree, the mangrove forests in this region support a variety of animal life, including more than 40 species of mammal and 200 species of bird. The mangroves are also essential as the nursery and feeding areas of many of the region's commercial fish and prawn species.

Because much of the human population of Southeast Asia lives in coastal zones, mangroves have been under particularly heavy pressure from overexploitation of their resources and from clearance for coastal aquaculture, agriculture and urban development. In the Philippines, for example, mangroves have been reduced in area by 75 percent during the last 60 years. In Malaysia, proposals for large-scale coastal reclamation threaten the remaining forests on the west coast of the peninsula. In Thailand, the area of mangrove is estimated to have been reduced by 22 percent, from 1,420 square miles (3,680 square kilometers) in 1961 to just under 1,120 square miles (2,900 square kilometers) in 1979. Clearance for aquaculture during the past decade has been especially rapid, and although figures are not yet available, it is feared that mangrove areas may have suffered a further

▼ *Flooded mangrove forest in Malaysia. Mangrove forests have been reduced on a vast scale in this region to make way for agriculture, aquaculture and urban developments. The cut mangroves are a much cheaper building material than expensive steel scaffolding. The large-scale destruction of mangroves in Southeast Asia has prompted some governments to introduce strict laws regulating the amount of mangrove forest that can be cleared.*

devastating decline. Fortunately, most of the mangroves' associated mud flats still remain, although some have been lost to aquaculture and industry.

Protecting the mangroves

In Indonesia, significant steps are now being taken to conserve the dwindling mangrove resource. The legal status of Indonesia's mangroves is such that commercial harvest of the trees is regulated and requires harvesters to leave an undisturbed protection zone 100 times wider than the tidal range along the seaward margin and 160 feet (50 meters) wide along rivers. Subsistence use of mangroves is not regulated, although major projects are currently being developed to improve management of mangrove forests by involving local communities in replanting and developing systems of sustainable use. More than 30 protected mangrove areas have been declared by the minister of agriculture and several others have been proposed. Similarly, in the Philippines a total of 300 square miles (780 square kilometers), 58 percent of the remaining mangrove forest, have been identified for designation as conservation and preservation areas.

And in Thailand, the government is providing funds for rehabilitation of mangroves in areas where aquaculture has proved unsuccessful.

Flooded forest

Peat swamp forest covers about 77,000 square miles (200,000 square kilometers) of Southeast Asia, mostly in Indonesia and Malaysia. Figures for freshwater forest are less comprehensive, but Indonesia alone possesses 20,000 square miles (50,000 square kilometers). Peat swamp forest normally develops from freshwater swamp forests in which leaf litter and other organic debris has accumulated in layers of peat up to 65 feet (20 meters) thick. The peat swamp is acidic, often domed and supports a forest that is characteristically zoned in concentric bands around the top central dome. The vegetation varies from about 100 tree species in the outer mixed-forest zone to virtually single species forest, known as *padang*, on the top of the domes.

Freshwater swamp forests and peat swamp forests play an important role in the mitigation of flooding in adjacent areas by acting as natural reservoirs that absorb and store excess water during the rainy season. These types of swamp are also major forestry resources, with many valuable timber species. The principal threats to

▼ *Farmers of the plains of northern Thailand transplanting upland rice seedlings. Many of southeast Asia's wetlands have been converted to ricefields. While this has resulted in the loss of biodiversity in many cases, many ricefields continue to provide important habitats for birds and other species of animal.*

▲ *Freshwater Swamp, Malaysia. Freshwater swamps are among the most threatened wetlands in Malaysia, with forested sites being particularly threatened. Clearance for agriculture and timber exploitation are the major threats.*

freshwater swamp forest are conversion to agricultural use and nonsustainable exploitation for timber. These problems are particularly acute in Malaysia, where freshwater swamp forest is probably the most severely threatened wetland habitat. Although peat swamp forest, unlike freshwater swamp forest, is of only marginal use for agriculture, clearance for aquaculture ponds is a major threat, together with nonsustainable logging. Under current practice, it is predicted that all of Sarawak's peat swamp forest will have been logged before the year 2000.

Berbak Nature Reserve

Berbak Nature Reserve in Jambi Province on the east coast of Sumatra covers an area of approximately 630 square miles (1,650 square kilometers). It consists mostly of peat swamp forest and contains an amazingly diverse flora. The forest also supports a remarkably diverse mammal fauna, including Sumatran tigers (*Panther tigris sumatrae*), clouded leopards (*Neofelis nebulosa*), Malayan tapirs (*Tapirus indicus*) and Sumatran rhinoceroses (*Dicerorhinus sumatrensis*). More than 160 species of bird are known to live in the reserve, and it may be one of the last remaining strongholds of the rare crocodilian, the false gavial (*Tomistoma schlegelii*).

The swamp forests of Berbak are thought to supply a wide range of benefits, including the support of fisheries, protection against saline inundation of inland agricultural areas and flood protection. The Asian Wetland Bureau and the Directorate General of Forest Protection and Nature Conservation of the Indonesian Government have recently completed a detailed study of the management needs of this area. The Berbak Nature Reserve is currently under severe threat from illegal logging and clearance, poaching and development. It was, however, declared as the country's first Ramsar Site when Indonesia ratified the Convention in 1992, and this should encourage greater support for conservation efforts in the area.

Australia

Covering 3 million square miles (7.7 million square kilometers), the Australian continent is vast, mostly flat and much of it extremely arid. Yet, when it rains, floods spread across large areas, contrasting vividly with the prevalent dryness. The flatness and the extremes of flood and drought are major factors in governing the distribution and type of wetland over much of the continent. Permanent lagoons, swamps and marshes lie in areas of higher rainfall, while in the arid zones wetlands may seem to be nonexistent for years, only to temporarily dominate the landscape after heavy rains. The flora and fauna of these wetlands are those that respond to episodes of plenty, and their ecology is overwhelmingly one of abundant life and widespread death.

Australia is, in geological terms, very old. Erosion and sedimentation have created extensive landscapes of extremely low relief; Lake Eyre, into which drains much of the central interior, is many meters below sea-level. Because of its geology, huge areas of inland Australia generate salts that accumulate in saline lakes which may be dry for years.

Australia's varied wetlands

Although essentially arid, some areas of Australia support a surprisingly rich variety of wetlands. Short, perennial streams drain across narrow coastal plains; large, biologically rich estuaries lie around the northern and eastern coastlines, while west of the Great Dividing Range, which almost joins Cape York Peninsula in the north to the southern margins of the continent, huge river systems drain into the interior. The Murray-Darling river system drains about 14 percent of the continent.

Wetlands lie along the courses of the drainage systems in the form of swamps, floodplains and stream beds. These are often dry for many months or years, but during floods they change dramatically, and parts of inland Australia may suddenly be awash with swirling, turbid water. Floodwaters spill out from the drainage channels, changing the parched landscape into a kaleidoscope of briefly flowering plants. After the floods, shallow inland lakes, which gradually recede over months and sometimes years, remain as oases teeming with fish, birds and marsupials feeding on the rich invertebrate and plant life. These wetland ecosystems are so significant that they influence the migrations of many birds, including shorebirds and seabirds that move inland in these times of plenty.

Map Note

Australia is a very dry continent but its interior is characterized by regions which flood periodically and become important wetlands. The reasons for the flooding and the periodicity determine the character of the wetland. Those which flood annually are shown as seasonally flooded and those which flood when they are subject to irregular rainfall are shown as saltpans/seasonal wetlands. A third category, shown as "occasionally flooded", are known locally as the Channel Country and depend on rainfall to the north and east which floods irregularly into an otherwise arid region.

Arafura Sea
Cobourg Peninsula
Timor Sea
Darwin
Kakadu National Park
Ashmore Reef National Nature Reserve
Ord River floodplain
Lakes Argyle and Kununurra
INDIAN OCEAN
Roebuck Bay
Eighty-mile Beach
NORTHERN
TERRITORY
Lake Mackay
Alice Springs
WESTERN AUSTRALIA
AUSTRALIA
hark ay
Lake Eyre
SOUTH AUSTRALIA
Perth
Peel-algorup system
Forrestdale and Thomsons Lakes
Becher Point Wetlands
Toolibin Lake
Vasse-Wonnerup system
Lake Gore
Lake Warden system
Muir-Byenup System
SOUTHERN OCEAN

0 km   500
0 miles   250
N

▓ Ramsar Sites
▓ Water Bodies
▓ Mangroves
▓ Salt Lakes

Wetland species

Where wetlands are more sustained or predictable, they support forests. In freshwater areas these forested wetlands are dominated by species of eucalyptus, such as the coolibah (*Eucalyptus microtheca*) and the river red gum (*Eucalyptus camldulensis*), paper-bark trees (*Melaleuca* spp.), she-oaks (*Casuarina* spp.) and the freshwater mangrove (*Barringtonia* spp.). On the floodplains of the Murray River, *Eucalyptus camaldulensis* grows in sufficiently large quantities to constitute a significant hardwood forest resource.

259

Both Northern Territory and Queensland have extensive freshwater floodplains covered by grasses, such as wild rice (*Oryza meridionalis*) and spiny mudgrass (*Pseudoraphis spinescens*), and tall trees, such as paperbarks. Reeds, sedges, water lilies and herbs abound in a diverse floral display.

The animal life of these seasonally flooded ecosystems is no less diverse, with the vertebrates being far better known than the myriad invertebrate species. Freshwater and saltwater crocodiles (*Crocodylus porosus* and *C. johnstoni*), turtles, such as the pignosed (*Carettochelys insculpta*), and fishes, such as the barramundi (*Lates calcarifer*), are numerous.

Ramsar Sites

Ramsar Sites/
Parks and Reserves

Water Bodies

Seaonally Flooded
Wetland

25–50% Wetland

Occasional Wetland

Mangroves

Peatlands

Salt Lakes

Saline wetlands

Mangroves and salt marshes are common along much of the long northern and western coastline, with the total area of mangrove in Australia estimated at over 4,500 square miles (11,600 square kilometers). Outside Indonesia, this is the largest area remaining in the countries of Southeast Asia and the Pacific. Pressure for coastal development has led to increasing loss of mangrove forest in many parts of Australia.

Growing awareness of their value is, however, leading to action. In Moreton Bay in southern Queensland, for example, it has been estimated that commercial and recreational fisheries based on mangrove-dependent species are worth about 280 million Australian dollars annually. This has resulted in the founding of a marine park, which in turn has halted the expansion of Gold Coast City, the main cause for mangrove destruction in Moreton Bay.

▼ *Eucalyptus*

Aboriginal use of wetlands

Long before Captain Cook set foot on Australian soil in 1770, the wetlands of northern Australia played a central role in the lives of many of Australia's aboriginal peoples, and they continue to do so today. Coinciding with the formation of the freshwater wetlands of northern Australia, between 500 and 1,200 years ago, there was a sudden increase in human occupation of the plains of the Alligator Rivers Region in the Northern Territory. People

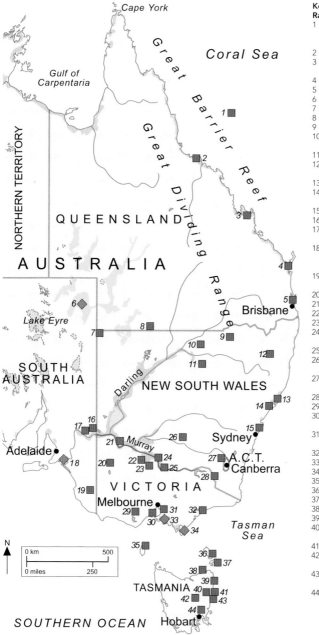

261

here are thought to have followed a pattern of seasonal nomadism, congregating on the wetland margins in the dry season and moving to higher ground in the wet season.

Today, the Australian Aborigines still use the food resources of wetlands, although the ready supply of food from feral buffalo has reduced the use of natural wetland foods. These buffaloes also degrade the wetlands and this has combined with saltwater intrusion, due to over-stretched groundwater reserves used for irrigation, and introduced-plant invasion to reduce the amount and number of types of food available.

Aboriginal women harvest plant foods from freshwater wetlands during the dry season, when the food is more easily accessible. Hunting and fishing are mainly done by the men.

▲ *Paper-bark trees* (Melaleuca leucadendron) *in wetlands.*

Lake Eyre Basin

Lake Eyre Basin is one of the largest internal drainage systems in the world. It covers 425,000 square miles (1.1 million square kilometers) and is underlain by the world's largest artesian basin. At its center, the Lake Eyre Basin is generally arid, with average annual rainfall between 4 and 6 inches (100 and 150 millimeters). The wetland systems the basin contains form oases in a vast stone and sand desert. The oases are among the most dynamic and spectacular in Australia. They are fed by the four major rivers that drain into Lake Eyre, namely the Cooper and Diamantina in the east, and the Neales and Macumba in the west.

Rainfall is erratic in timing, location and intensity, so the discharge of even the large rivers is extremely unpredictable. Yet it is this variability, combined with the low grades and the relief of the basin, that is responsible for the complexity of the Lake Eyre wetlands. Plants and animals respond with opportunism to the instability of the wetland environment. Plants flower and fish reproduce after floods, and waterbirds congregate to feed and breed, generating some of the highest waterfowl densities in the world. In droughts, wetland animals emigrate, aestivate (lie dormant) or, like fish in the contracting lakes, die in multitudes. In such an environment, the permanent wetlands are vital refuges for the less mobile species. The mound springs in the south of the basin are fed by artesian waters, and provide islands of permanence; and inland Galápagos, they are rich in unique species.

Growing threats

Because the rivers have not been substantially modified or polluted, the wetlands are still relatively pristine, and no significant wetland loss has occurred. However, this situation is now changing. The whole basin is subject to intensive grazing pressure from rabbits and cattle and much of it is being dissected by mining exploration tracks. In addition, the acacia woodlands of the headwaters are being cleared, and damage from four-wheel-drive tourism is increasing rapidly.

Kakadu National Park

A report to Unesco's (United Nations Educational, Scientific and Cultural Organization) World Heritage Committee stated that: "There is simply no other protected area on Earth like Kakadu." Stretching over 7,700 square miles (20,000 square kilometers) and including almost the entire catchment area of the South Alligator River, the park is the largest in Australia and includes a tremendous diversity of habitats and species found nowhere else in the world. Located in the north of the Northern Territory, Kakadu's wetlands are listed under the Ramsar Convention, and include mangroves on offshore islands, along the coast and fringing the rivers, salt flats, and creeks that overflow onto extensive freshwater wetlands, grasslands, sedgelands and swamps. The biological value of these wetlands is enormous. The vegetation is species-rich with, for example, 22 mangrove species and 225 freshwater plants on the Magela floodplain. The birdlife is prolific, with peak populations reaching 3 million individuals, 85 percent of them magpie geese (*Anseranas semipalmata*). Up to 400,000 wandering and 70,000 plumed whistling ducks (*Dendrocygna eytoni*), 20,000 Radjah shelducks (*Tadorna radjah*), 50,000 Pacific black ducks (*Anas superciliosa*) and 50,000 gray teal also use

▼ *Gurra gurra wetlands in the lower River Murray, recent beneficiary of an effective desalination project.*

the area. Thirty-five species of shorebird, many of them migrants from the Arctic, use these wetlands as well. The wetlands near the coast support turtles and the estuarine crocodile. Other turtles, snakes and the freshwater crocodile also live in the freshwater wetlands, along with seemingly innumerable frogs.

Change and development

Management problems center on introduced animals and plants, tourism and mining operations. Buffalo were a major problem on the freshwater floodplains, but have been virtually eliminated. Feral pigs, however, are a serious problem and extraordinarily difficult to control. As for vegetation, the most serious threat is posed by the prickly mimosa shrub. To date, it has been kept at bay in the park, but severe infestations on the park's boundaries make this species a potential problem. Increasing numbers of tourists have been visiting Kakadu over the last decade. Balancing the needs and expectations of these visitors with nature conservation priorities is an ongoing process. The issue that has generated most concern and controversy is the development of uranium mining in the Magela Creek catchment area.

New Zealand and the Pacific

The volcanic origins and glaciated history of New Zealand combine with today's maritime climate and heavy rainfall to produce a diverse wetland landscape. Features include such things as rivers and bogs from frequent rain, glaciation and volcanic action, swamps from the deposition of erosion products by rivers and the sea, and estuaries and lagoons from tidal flooding of old Pleistocene (Ice Age) valleys.

The wetlands are widespread and diverse. They support distinctive communities, which contribute to the unique biological and geographic character of New Zealand. They include swamps of *Phormium* spp., braided rivers, which form good waterfowl habitat, bogs, saline rush and reed estuaries with *Leptocarpus* spp. and *Juncus* spp., and kahikatea (*Podocarpus dacrydioides*) swamp forest.

The distribution of these wetlands reflects a combination of geological history, relief, climate and the intensity of wetland modification. For example, freshwater wetlands

▼ *Blue River, New Caledonia. The dominant wetlands of the Pacific are shallow coastal waters and lagoons, together with associated coral reefs. These wetlands play a critically important role in the livelihoods of local people and support important biodiversity.*

occupy only a few square kilometers in the eastern North Island, while in South Island they occupy some 120 square miles (300 square kilometers). While of the 300 or so estuarine wetlands distributed around the country's coastline, some of the largest systems occur in the northern end of North Island.

Conversion and loss

There is a long history of use and modification of wetlands in New Zealand. Land drainage, gold mining, flood control, land clearance, general agricultural development, kauri-gum digging and flax milling have all contributed to wetland loss. It is estimated that only 10 percent of pre-European wetlands remain, and wetlands now occupy less than 2 percent of the total land area of New Zealand.

The process of wetland loss has fragmented the wetland resource, and conservation action needs to be increased in the face of continuing pressure from extraction of sand and gravel, reclamation of estuaries, lagoons, lake shores and river margins, flood control, pollution and land drainage. Priorities include the establishment of buffers of indigenous vegetation along rivers and around the margins of lakes, swamps and estuaries, and the protection of corridors linking wetlands of all kinds to other protected ecosystems, both terrestrial and marine.

The Pacific islands

Perceived by many as paradise, the total land area of the 23 small island nations and territories of the Pacific region is merely 41,000 square miles (107,000 square kilometers), rather less than that of North Island, New Zealand. Thus, while these islands support a wide variety of wetland types, most are very limited in their extent because of the tiny size of most of the islands. Yet, despite their small size, these wetlands yield a variety of resources, some of which are used intensively by the island peoples. Many islands, for example, provide the basis for sustainable *taro* agriculture, a crop grown for its edible root.

By far the most extensive wetlands are the shallow lagoons and reef flats that fringe many of the larger islands and comprise the bulk of most of the low-lying atolls (ring-shaped coral reefs surrounding a central lagoon and with breaches to the open sea). Mangroves also occur widely in the western and northwestern Pacific region.

Freshwater wetlands, such as lakes, marshes, swamps and bogs, are well represented on the larger islands of New Caledonia, Fiji and the Solomon Islands, but are very limited on the smaller islands, and virtually nonexistent on atolls. Most of the freshwater lakes lie on mountainous volcanic islands, where they have formed in extinct or dormant volcanic craters. Examples include Rano Kau, Rano Raraku and Rano Aroi on Easter Island, Lake Waiau on the Big Island in Hawaii, Lagona Lake and Inner Lake on Pagan in the northern Marianas, the Lakes of Niuafo'ou and Tofua in Tonga, Lake Letas in the Banks Islands (the largest freshwater lake in the Pacific islands) and Lanioto'o Lake on Upolu in Western Samoa. Many of these lakes have special cultural importance for the local people.

Extensive areas of swamp or marsh are rare on the Pacific islands, notable exceptions being Kawainui Marsh on O'ahu (the largest freshwater swamp in Hawaii) and the Plaine des Lacs, a complex of freshwater lakes and marshes covering 75 square miles (200 square kilometers) near the east end of the island of New Caledonia. Although extensive areas are rare, there are small areas of freshwater marsh and swamp on most of the larger islands. Typically, these wetlands are fringed with tall stands of the reed *Phragmites karka.* Almost wherever they occur, these wetlands have been modified for the production of *taro.* Extensive *Sphagnum* bogs are confined to high elevations, usually over 5,000 feet (1,500 meters) on the larger Hawaiian Islands, such as Alakai Swamp on Maui.

Human impact

In a region where flat land suitable for agriculture and development is very much at a premium, wetlands in the Pacific are often seen as wastelands, easily converted to other uses. This situation is further exacerbated by the rapid growth in population. Because of their small size, the freshwater wetlands of the Pacific are extremely vulnerable to modification or

PACIFIC
OCEAN

Tasman Sea

Auckland •
Whangamarino ■ ■ Firth of Thames
Kopuatai
Peat Dome

NEW ZEALAND
NORTH ISLAND

Hawke's Bay

• WELLINGTON

■ Ramsar Sites

destruction, and many natural marshes have now been drained for agriculture and other forms of development.

Coastal wetlands and mangroves have frequently been modified or destroyed by infilling to provide land for housing, industry, airports and harbors. Similarly, because of the restricted development of mangrove swamps in the Pacific and their often stunted growth forms, the value of mangroves has not been widely appreciated and little legislation exists for their protection. In Fiji, large areas of mangrove swamp have been destroyed for shrimp production and the cultivation of sugarcane and rice, while in Western Samoa, mangroves are being reclaimed for settlement or used as rubbish dumps.

Polluted waters

Massive amounts of sediment, transported downstream as a result of poorly managed forestry, slash-and-burn agriculture and mining activities, have led to the siltation and destruction of many coastal wetlands. Seagrass beds and coral reefs near the shore are particularly affected by this problem. The discharge of urban and industrial effluents has caused serious pollution in wetlands and coastal waters, and wetlands have all too frequently been used for the disposal of rubbish and waste oils. On many of the more densely populated islands, especially in Micronesia, high levels of pollution have resulted in the

---

■ Farewell Spit

Farewell Spit is a unique, 18-mile (30-km) long sandspit lying at the north of Golden Bay, South Island. It covers a land area of some 8 sq mi (20 sq km). The spit is estimated to have originated 6,500 years ago, derived from material eroded from the Southern Alps and West Coast seacliffs. Because the southern end of the spit has extensive tidal sand and mud flats it provides

feeding areas or habitats for 83 species of wetland bird, both national and international waders, including knots, eastern bar-tailed godwit, turnstone and South Island pied oystercatcher (*Haematopus ostralegus finschi*) and is a major overwintering area for an estimated 12,000 black swans. In 1976, the spit was listed under the Ramsar Convention.

mass extinction of stream floras and faunas. The use in prawn fishing of chlorox to poison streams has been particularly harmful on some islands. In French Polynesia, the main rivers on Tahiti (Papenoo and Punaruu) are now badly degraded by mining for sand and gravel.

Despite the diversity of these pressures, their impact is often only localized, and there are fine examples of natural wetlands on many of the larger islands. Most of the governments in the region are becoming more aware of the need to use these valuable resources wisely, and in several countries, coastal zone management plans have been developed which take into consideration the need for wetland conservation.

Lakes of the Pacific region

Because of their extreme isolation, many of the freshwater lakes, marshes and streams in the Pacific islands have various rare and endemic species of flora and fauna. Streams in American Samoa and on several of the Micronesian islands, for example, are home to fishes that are found nowhere else in the world. There is at least one endemic species of fish, *Neogallaxia neocaldonicus*, confined to Lac en Huit in the Plaine des Lacs in New Caledonia, and an endemic sea krait, *Laticauda cruckeri*, is confined to Lake Te-Nggano on Rennell Island in the Solomons.

About 17 species of waterbird are endemic to the Pacific islands, but seven of these are now believed to be extinct (the Mariana mallard, four species of rail and crake, the Samoan woodhen and the Tahitian sandpiper), and all the others are under threat. In some cases the threat to species is a direct result of wetland destruction, but in others it is primarily caused by introduced predators. One species, the Laysan teal, is confined to a single brackish pond on Laysan Island in the northwest chain of the Hawaiian islands.

Endemic plants include a species

▼ *Cushion plant (Donatia novae-zelandiae) community at Awarua Bog, a blanket bog that forms part of the Waituna Lagoon Ramsar site, New Zealand.*

of tree in the swamp forests of Yap and Palau, and a palm, the ivory nut palm, in the swamp forests of Pohnpei and Chuuk in Micronesia.

Kopuatai Peat Dome

The Kopuatai Peat Dome, a Ramsar Site on the Ilauraki Plain, covers 3,730 square miles (9,665 square kilometers) of the largest raised bog left intact in New Zealand. It is also the largest lowland bog dominated by the giant "restaid" rush, the greater wire rush *Sporodanthus traversii*, and the only significant unaltered restaid bog in the world. The peat began developing 13,500 years ago, and at present the depth of the peat is up to 40 feet (12 meters) toward the center. The hydrological regime of the peat is dominated by rainfall, receiving little nutrient-rich groundwater. As a result, the peat is acidic and low in nutrients. In contrast, in the surrounding swamplands, where occasional flooding leads to mineralization of the soil, nutrient levels are much higher and the system more productive.

▲ *Hawaiian Moorhen at Hanalei National Wildlife Refuge, Kilauea, Kauai, Hawaii. The Hawaiian Moorhen is just one of the many endangered species found here. Visitors can get a closer look at this and other endangered species (Hawaiian coot, Hawaiian stilt, Hawaiian duck) by renting kayaks and traveling on the Hanalei River.*

Nine species of threatened plants and animals depend on the dome, including the endemic black mudfish (*Neochanna diversus*), the Australasian bittern (*Botaurus poiciloprilus*), North Island fernbird (*Bowdleria punctata vealeae*), banded rail (*Rallus philippensis assimilis*), marsh crake (*Porzana pusilla affinis*) and spotless crake (*Porzana tabuensis plumbea*). Threatened plants that find a refuge on the dome include the endemic greater jointed rush (*Sporadanthus traversii*), which covers 8 square miles (22 square kilometers), three bladderworts, the creeping club moss and a fern.

Because of the fragile nature of the patland ecosystem, entry without a permit is prohibited. However, gamebird hunting, grazing and a small area of agriculture are allowed in the mineralized fringe. Expansion of agriculture is at present the most severe threat to this fringe.

Visiting Wetlands

WHERE TO GO?

The Pantanal in South America or Botswana's Okavango Delta exemplify the spectacular, globally important, wetlands that provide "once in a lifetime" experiences for those who have the opportunity to visit them. Yet as this Guide shows, wetlands are some of the world's most widespread ecosystems and are used by hundreds of millions of people each year. For the majority of these people in the developing world wetlands provide a source of food and income, but for millions of others in both developed and developing countries wetlands are a place for leisure and education, and a source of enjoyment. Because they are so diverse and rich in wildlife, yet so accessible, they are one of the most important ways for people to experience nature and enjoy its diversity.

For those who live on the coast, estuaries, salt marshes and other tidal wetlands are just a short journey away. Whether these are in national parks, nature reserves, or other protected areas, they provide opportunities to watch birds or simply enjoy the spectacle of changing light and seasons in the natural landscape. Similarly for those who live inland, but within easy reach of lakes, rivers, or valley bottoms, many freshwater wetlands are easily accessible and provide opportunities to see birds and other wildlife even on short weekend or evening walks. One of the best ways to get to know wetlands and appreciate their biological importance is to take a break from a walk and sit down on their edge. After a few minutes, we begin to notice the birds that alight on the reeds, or hear the frogs as they begin to call. In the mornings and evenings the lucky ones among us might see a fox or an owl that has come to the wetland to hunt small mammals and birds that find shelter in the reeds and grasses on the wetland edge.

The educational and leisure opportunities of wetlands have been developed through visitor facilities at many protected wetlands. These include education centers with displays for the public and more elaborate materials for school visits and training courses held at the site. Others include guided tours by foot, boat, or even small trains and horseback, along trails that can bring visitors into close contact with the diverse wetland habitats and many of their most abundant species. Many wetland sites have raised wooden trails that take the visitor across the wetland and provide a "birds-eye view" of many habitats including open water. Hides or blinds for watching wildlife are also common and provide great spots for getting to know the daily and seasonal rhythm of the wetlands.

Read more on visitor centers through the web sites listed in the Resources section.

WHAT TO DO?
Bird watching

Birds are the most spectacular and accessible wildlife in most of the world's wetlands. Because wetlands are so productive they support waterbirds in massive numbers, ranging from the 12 million individuals of over 50 species that come to the Wadden Sea coast of the Netherlands, Germany and Denmark, to the one million Lesser Flamingoes that use Kenya's Lake Nakuru. But even small wetlands provide important opportunities to watch birds, often at relatively close range and this has made wetlands amongst the most preferred habitats for bird watchers all over the world.

▼ A birdwatcher observes snow geese from a viewing point at Chincoteague National Wildlife Refuge, Virginia, USA.

In most wetlands the opportunities to watch birds vary with the season and time of day. In the northern hemisphere the winter months, and the autumn and spring migration seasons are the best times to visit estuaries and other coastal wetlands. Here birds are dispersed at

low tide, feeding on the invertebrate food supply hiding in the mudflats. When the tide turns however the birds slowly move inland running or flying to escape the rising waters. At high tide vast numbers of shorebirds congregate on secure roost sites, and when disturbed by predators wheel majestically along the coast like a swarm of tightly packed aerial acrobats. Some of these migrating shorebirds cross the equator and spend the northern winter in the southern summer. Large concentrations in particular can be found in parts of Argentina, South Africa, and Australia.

The winter and migration seasons are also the best time to watch concentrations of geese and ducks in the northern hemisphere. One of the characteristic sights of the winter months in many regions is the skeins of geese moving between wetland areas and to and from their feeding sites. In Europe some of the most famous sites for geese are the islands and surrounding intertidal flats of the Wadden Sea, and the estuaries and mudflats of the British Isles. Swans also migrate south and concentrate in selected sites. One of these in Britain is the headquarters of the Wetlands and Wildfowl Trust at Slimbridge where the same individually marked Bewick Swans return year after year after having spent the breeding season on the Siberian tundra. These migrations are mirrored in North America, where tens of millions of ducks and geese migrate from their summer breeding sites in the Arctic and midwest to winter along the southern rivers and coasts.

In summer these same wetlands can seem empty, with only a few resident birds, or late migrants occupying the marshes and mudflats. However to the careful observer there is still much to see, and many people who live near these sites track the changing wildlife as the seasons cycle. During the early fall and spring in particular bird populations can change on a daily basis, rather like a hotel that is always full but where guests arrive and depart each day. Studies of these migratory populations have shown just how complex their use of wetlands can be and how the survival of many of these populations is dependent on a range of sites along their migration routes and in their wintering areas.

Other wildlife

While birds are the dominant feature of wetland wildlife in most parts of the world, they are overshadowed in some by spectacular mammals and reptiles. The larger

▲ *Bombay Hook National Wildlife Refuge, Delaware, USA. Hundreds of thousands of ducks and geese arrive in the autumn to rest and feed before heading away for the winter. In spring, millions of horseshoe crabs come ashore to lay their eggs, providing food for tens of thousands of shorebirds.*

wetlands of Latin America – the Pantanal and the Llanos – are famous for the large numbers of Capybara that graze the wetland vegetation and the even larger numbers of caiman. These are both easily accessible for visitors and provide striking photo opportunities. The Pantanal is also one of the strongholds of the jaguar and a place where they are relatively easily observed. Time spent watching for jaguars here is often rewarded with a view of these spectacular predators.

In Africa the floodplain wetlands of eastern and southern Africa support large populations of antelope, of which several species of lechwe are the most abundant, and the sitatunga the rarest. To see large herds of lechwe splashing through the shallows of the Okavango in Botswana, or Kafue Flats and Bangweulu floodplain in Zambia, is one of the most spectacular wildlife experiences in these areas. Together with elephants and hippos, these antelope herds thrive on the productive wetland vegetation, and provide a stunning wildlife spectacle that draws thousands of visitors each year. This is especially true of the Okavango which is larger, and has good populations of lions, leopard and other predators that thrive on the antelope, and where the infrastructure for hosting visitors is better developed.

Fishing and hunting

Every year anglers spend billions of dollars for the right to fish in rivers, lakes and other wetlands. As far back as 1969 it was estimated that in California each salmon cost an angler over US$18. Today angling is one of the most popular outdoor sports and has become increasingly international with dedicated anglers traveling all over the globe to fish new species in new environments. In the USA alone 34 million people went fishing in 2001, and spent a total of US$ 36 billion; and there are many millions of anglers in Europe and other regions. Angling vacations are increasingly popular, with specialty fish such as tarpon in Central America, tiger fish in southern Africa, and Nile Perch in Egypt's lake Nasser attracting increasing numbers of enthusiasts each year.

Waterfowling is also a very popular sport in many parts

▼ Combat fishing for Chinook (King) Salmon in a southcentral Alaskan stream.

▲ *Canoeing at Bond Swamp National Wildlife Refuge, Georgia, USA.*

of the world, with some 3 million waterfowl hunters in the USA in 2001. While the enthusiasm for hunting has caused conflict with conservation in some countries such as France, it is widely recognized in many others that hunting and conservation have important synergistic roles. In North America, Ducks Unlimited originated in the hunting community, but is widely regarded to have played a central role in the conservation of the continent's waterbirds. Many of the world's most respected conservationists started as hunters, with Sir Peter Scott, the founder of the Wildfowl and Wetlands Trust and one of the founders of WWF, being one of the best examples.

Boating and Canoeing

One of the best ways to see wetlands, and to enjoy the sense of wilderness that is possible in even some relatively small sites, is to take a boat or canoe through the wetland channels. In North America canoeing is a popular pastime in many wetlands including the Everglades where extended canoe trips are possible. Similarly in the developing world organized boat trips are now a common feature in many of the larger wetland systems. For example in the Llanos of Venezuela or the Pantanal in Brazil small boats provide a magical opportunity to visit the backwaters of these wilderness areas and surprise turtles basking on logs or caiman sleeping on the shore. In Africa extended canoe safaris are possible along the middle stretches of the Zambezi river, providing an opportunity for visitors to get unforgettable water-level views of hippos and other wetland wildlife. Travel by makoro (traditional canoe dug-out made from ebony or sausage tree logs) is also one of the best ways to visit the Okavango Delta in Botswana. While these are far from the luxury of some of the larger tented camps, they provide an intimate introduction to the delta

Wetland education

Because wetlands are so readily accessible to many of us, and because they are frequently the last remaining "wild" spaces in an urban and agricultural landscape, they provide wonderful opportunities to educate about the natural environment, how ecosystems function, the benefits they bring, and the conservation challenges being faced. This has been recognized increasingly over the past 30 years or so and there are now a wide range of educational tools and initiatives that build understanding of wetlands, and through this build understanding of a wider set of environmental and conservation issues.

For schools wetlands provide a particularly rich set of opportunities to learn: about the formation of different landscapes and the wetlands within them; about the history of human use of wetlands and their role in early and modern societies; of the role of wetlands in filtering and storing water; of their contribution to flood control and the consequences of wetland loss; of the biological complexity of the natural landscape around us; of the seasonal cycles of both plants and animals and behavior of a wide range of birds, mammals, amphibians, reptiles and insects; and of the complex decisions that are required to achieve conservation in modern societies. There is even the possibility of cooking classes that use wetland produce, starting with the more obvious fish and game from wetlands, but also including a wide range of wetland plants that can serve as staples (such as rice or sago) or as relish such as watercress and blueberries.

▼ Environmental Education at Lake Woodruff National Wildlife Refuge, Florida, USA.

TWENTY TO VISIT

One of the messages of this Guide is that wetlands are widely distributed across much of the world and are relatively accessible to most of us on a daily basis. However the Guide also shows that some of the larger wetlands provide some of the most spectacular wildlife viewing experiences on the planet, while also continuing in many to support flourishing local economies and cultures. At some stage we hope that readers will have the opportunity to visit some of these larger sites. To help decide where to go here are some suggestions.

AFRICA

Inner Niger Delta

Lying in central Mali on the edge of the Sahara desert, the inner Niger Delta provides a unique combination of herding, fishing and farming communities together with spectacular concentrations of breeding and wintering waterbirds. Accessible from the town of Mopti on the Bani tributary of the Niger, visitors can hire wooden canoes (piroques) to travel out into the delta. Alternatively for the more adventurous a journey on one of the larger canoes (pinasses) that are used to transport people and cargo through the delta can take travelers to Timbuctu, some 300km downstream. The best time to go is between November and February, the cool season when the river is receding from the floodplain, fishing and cattle grazing is intense, and the bird numbers are highest.

Banc d'Arguin

Lying on the coast of Mauritania, the Banc d'Arguin is the most important area of intertidal flats on the coast of Africa. In winter it is home to over 2 million shorebirds, while some 40,000 waterbirds, including Great White Pelicans, Greater Flamingoes, Spoonbills, and several species of tern also breed here. The Banc d'Arguin is also an important fishing area and the Imraguen fishing community and their predecessors have lived from the local fishery for the past 500 years or more. The sight of their lateen sailed boats working their way up the channels between the mudflats, backed by the dunes of the Sahara, is a unique spectacle in Africa. The Banc d'Arguin is accessible from the capital Nouakchott 150km to the south. The best time to visit is October to February when migratory bird populations are at their highest and when the Imraguen fish for migrating mullet.

Bijagos archipelago

A mosaic of palm covered islands, mangroves and inter-tidal mudflats, the Bijagos Arhipelago is another unique coastal wetland site lying off the coast of Guinea Bissau. Like the Banc d'Arguin it supports important populations of shorebirds during the European winter, where these mix with hippos, otters and manatees. The archiplago is accessible through the national capital Bissau using the local ferries that service the islands.

Bangweulu

The Bangweulu basin lies in northern Zambia and is a unique mix of lake, papyrus swamp and floodplain. It supports important populations of antelope, especially black lechwe and sitatunga, as well as wattled cranes and the rare shoebill stork. Bangweulu is rather isolated and less well know than some of the other wetlands of south-ern Africa but retains a pristine character that is well worth the visit. Safaris to Bangweulu can be organized through most tourist companies in Lusaka or Livingstone.

Okavango

Regarded by many as the jewel of Africa's wetlands, the Okavango is a spectacular mosaic of woodland and grassland savannah, floodplain and papyrus swamp that

▼ *Red lechwe* (Kobus leche), *running across the floodplains of the Okavango Delta, Botswana.*

▶ *The shoebill stork (Balaeniceps rex) is one of the most exotic of Africa's wetland birds. largely confined to Papyrus swamps the shoebill has strongholds in Bangweulu in Zambia and Lake Kyoga in Uganda.*

supports one of the most important concentrations of wildlife in Africa. For this reason it is also one of the most visited and many safaris have become extremely expensive. However there is a wide range of options and the visit is worth the extra expense. Safaris can be organized through many companies in Botswana and internationally, and should include time spent traveling in a mokoro (a local canoe). The best time to visit is July-September when the water levels are at their highest. As the waters recede however larger concentrations of birds and other wildlife can be seen.

AMERICAS
Pantanal

Often called the world's largest wetland, the Pantanal covers 54,000 square miles (140,000 square kilometers) and supports a rich and varied wildlife. Amongst these the jaguar is the most important and the open habitats of the Pantanal make it one of the best places to see this rare cat. Giant Otters also occur in good numbers and are seen frequently. Together with the large populations of caiman and capybara and the abundant waterbird populations the Pantanal and the Llanos are the only wetland sites in Latin America that can compare to the larger floodplains of Africa. Ecotourism is growing in the Pantanal and there are now many opportunities for visitors to stay on ranches in rich wildlife areas.

Llanos

Though smaller than the Pantanal, the Llanos is still massive, stretching over 40,000 squar miles (100,000 square kilometers). It is home to important populations of caiman and capybara, and large concentrations of waterbirds. Boat trips through the smaller rivers flowing through the Llanos are a particularly good way to see the wildlife of the area. As in the Pantanal there are good opportunities to stay at ranches in the Llanos. Visits can be organized through tour groups in Caracas.

Amazon floodplain

The Amazon tends to be synonymous with tropical forest, yet along much of its length the Amazon supports extensive wetlands that are some of the most important

▼ Giant lily pads on the Amazon. For visitors to Manaus a boat trip along the Amazon can include side trips into the floodplain where these giant water lilies provide an accessible and wonderful spectacle.

▲ *Lake Titicaca. Famous as the world's highest navigable lake, Titicaca provides a unique experience for visitors to Peru and Bolivia.*

on the continent. From the region around Manaus at the confluence of the Rio Negro and Solimoes rivers to that around Belem at the coast, the river supports extensive floodplain forests that yield a vast array of products to the people of the river and sustain extensive fisheries. For the adventurous boat trips can be arranged between Manaus and Belem, or shorter excursions organized to and from either city. One of the highlights of a trip from Manaus is the confluence of the milky muddy waters of the Solimoes and the dark organic water of the Rio Negro where both species of Amazon river dolphin can be seen.

Lake Titicaca

High in the Andes, Lake Titicaca is a spectacular landscape. While it does not provide the wildlife spectacle of the other wetlands listed here, the combination of the deep blue lake bordered by the dry soils of the Altiplano and framed by the snow-clad peaks of the high mountains, makes Titicaca a special place for many travelers. In this environment the local communities who continue to

▲ *Fishing boats on*
Tonle Sap

depend upon the lake live on floating islands in the lake and use reeds to build their homes and boats. There is nowhere else like it. Trips to Lake Titicaca are best organized from Lima in Peru.

Everglades

The Everglades have become synonymous with much that is special about wetlands in the United States, and equally serve as a symbol of the conservation challenges they face. On Miami's doorstep they are also one of the most accessible large wetland systems in the world and receives over one million visitors each year. The visitor facilities provide great opportunities to learn about wetlands and the many raised walkways give easy access to the complexity of the wetland ecosystems there. For those with more time a canoe trip through the wetlands is recommended.

Chesapeake Bay

Lying only an hour or so by car from the nation's capital, Chesapeake Bay brings internationally important wetlands to the doorstep of millions of Americans. Every winter it supports about one million waterbirds, and is an important recreational area throughout the year. Fish and shellfish from the Bay are also famous and a major attraction for tourists and locals. The Bay can be visited easily in a day or weekend outing from Washington or any of the other cities in the area.

ASIA
Tonle Sap
The Great lake of the Mekong "Tonle Sap" has lain at the heart of Khmer civilization for much of the past 2000 years. Today it continues to support a thriving fishery that plays a central role in feeding the people of Cambodia and in sustaining the country's economy. A visit to Tonle Sap therefore provides both a unique natural spectacle and an insight into how societies continue today to depend on natural wetland ecosystems. Boat trips on Tonle Sap can be organized from Siem Reap on the margins of visits to the temples of Angkor. The lake level falls as the Mekong flood waters recede. Best times to visit are in December–February when the flood levels are high.

Poyang Lake
Lying in the floodplain of the Yangtze valley, Poyang Lake (Poyang Hu) is one of the largest lakes in China. During the winter months it supports important populations of

▶ *Roseate Spoonbill (Ajaia ajaja), Everglades. The waterbirds that feed and breed in the Everglades are the most visible examples of the biodiversity that has made this unique wetland so famous.*

▲ *Yellow Water Lagoon, Kakadu. The freshwater lagoons are some of Kakadu's most beautiful wetlands. During the dry season large numbers of birds concentrate here.*

cranes and other waterbirds, notably most of the world population of the Siberian crane. With the construction of the Three Gorges Dam the long term future of Poyang Lake is of concern and it is uncertain whether it will continue to support such rich biodiversity. Trips to Poyang can be organized from Shanghai.

Sundarbans

The Sundarbans support the world's single largest stand of mangroves. These in turn provide a refuge for one of the most important remaining populations of Bengal Tigers, and a diversity of other wildlife. To travel by boat through the Sundarbans is a special journey into a land where people visit but nature continues to dominate. Visits to the Sundarbans can be organized through the Bangladeshi capital Dhaka and in the Indian Sundarbans through Calcutta.

Hokkaido marshes

In the mountainous landscape of Japan's largest northerly island, the reedbeds of Kushiro form an extensive wetland system that are both beautiful and of great biological importance. These reedbeds are the breeding site of Hokkaido's famous red-crowned cranes which have made the marshes internationally famous. Whether visited in spring and summer when the marshes are flushed green with new growth, or in winter when the straw colored reeds emerge from ice and snow, Kushiro is a special place. The highlight however is February and March when the cranes dance their courtship display on the snow-covered landscape.

AUSTRALIA
Kakadu

Lying a few hours drive east of Darwin in the Northern Territory, Kakadu is Australia's flagship wetland site. Covering 7,700 square miles (20,000 square kilometers) the National Park is a mosaic of freshwater and saltwater wetlands that support a great diversity of plant and animal species, including some 3 million waterbirds. Kakadu also occupies an important place in the lives of the local Aboriginal community with many sacred sites in the wetland and on its margins, and several important rock art sites within the National Park. Kakadu can be reached easily from Darwin and visitors can stay in hotel facilities within the Park.

▼ *Adult red-crowned crane (Grus japanensis) guard calling. This haunting bugle-like call evokes the beauty of Hokkaido's marshes.*

EUROPE
European estuaries

The string of estuaries dotted along the coast of western Europe are some of the most important intertidal wetlands in the world. The largest of these is the Wadden Sea along the coast of the Netherlands, Germany and Denmark, but France, Belgium, Ireland, and the United Kingdom all have sites that are of considerable international importance. Together these sites support many millions of waterbirds during the course of the year and provide a wonderful natural spectacle for the people who live nearby or visit on weekends and on holiday. They provide the people who visit with a hint of the natural spectacle that is possible in many other more distant wetlands and this has done much to build international support for wetland conservation.

Danube Delta

The Danube is the second longest river in Europe and the delta is one of the largest wetlands on the continent. Covering 2,000 square miles (5,200 square kilometers) the delta supports a diversity of wetland habitats and a

▼ *Mont St Michel, France. At low tide the bay around Mont St Michel is one of the most important estuarine wetlands in northern France, and one of the country's 22 Ramsar sites. During autumn and winter these mudflats provide important feeding grounds for migratory waterbirds. Over 100,000 shorebirds winter in the bay.*

rich waterbird population. For many years it was very difficult to travel to the delta, but with the opening up of Romania to the European Union it is now relatively easy to visit the delta and enjoy the wilderness that it still supports. Boat trips through the delta's waterways are the best way to see the area.

Camargue

For many years the Camargue has served as a symbol of wetland conservation in the Mediterranean. Easily accessible for millions of Europeans and providing spectacular views of the greater flamingoes and other waterbirds that breed and winter there, the Camargue has helped build awareness of the importance of the region's wetlands and of the challenges they face. Visitors to the Camargue can obtain information from several information centers, and can watch birds from the network of roads running through the area and from many vantage points looking our over the marshes. There are also several areas where trips by horseback are available.

Coto Donana

In the early 1960s the protection of the Coto Donana was one of the first major successes of the fledgling World Wildlife Fund. The area continues to be Spain's most important protected area and is on a par with the Camargue in its international importance. Once the refuge of only the kings and queens of Spain, guided visits to the Park are now possible through the Park Visitor Center. This is located near the town of El Rocio whose annual festival links the local culture closely to the wetland.

Wetland Conservation

This guide celebrates the diversity and importance of the world's wetlands. However it also highlights the pressures upon them and the need for concerted action if wetlands are to be sustained for future generations.

At the start of the 21st century there is both good and bad news for wetlands. On the positive side more attention is being given today to wetlands conservation than at any time in the past. The rise in signatories to the Ramsar Convention, from 23 at the start of 1980 to 55 at the beginning of 1990, and 144 at the start of 2005, is but one indication of this success. The number of wetlands protected under the Convention, now standing at over 1400, and covering over 295 million acres (120 million hectares), is another. Arguably more important are the number of countries that have developed, or are working to develop, national wetlands policies and other related conservation programs.

A large number of non-governmental organizations are also investing in wetland conservation and at international level some of the biggest conservation groups give high priority to wetlands. IUCN – The World Conservation Union, and the Worldwide Fund for Nature (WWF) have the largest investments in this work, but other major contributions are made by Wetlands International, Birdlife International, the International Crane Foundation, the International Rivers Network, and many others.

On the other hand pressures on wetlands are still substantial and many important wetlands may not

▼ *A family of black-necked cranes* (Grus nigricollis) *in Cao Hai Nature Reserve, China. The International Crane Foundation, in co-operation with the Guizhou Environmental Protection Bureau and the New York based Trickle Up Program, started a model program at the Cao Hai NR which provides grants as incentives to businesses that will be less detrimental to the reserve's wetlands; promotes reforestation and will develop management plans for the reserve.*

▲ *Fall hues begin to appear on trees and tundra at the Togiak Refuge, Alsaka. One of the likely consequences of global warming is an extension of the boreal forest northward into what is now tundra. The encroachment by the forest will engulf many waterbird breeding sites, and populations could decline dramatically.*

survive. The guide provides many examples of these pressures, from canal construction affecting the Danube delta, dams on the Mekong, Yangtze and other rivers, oil exploration in West Africa and the Arctic, and drainage for agriculture and urban development in most regions. These examples reflect the continuing pressures that economic development brings for the natural environment, and those who depend most closely upon natural ecosystems. It is to mitigate these risks, and reconcile the need for economic development with wetland conservation and sustainable use of their resources, that major investments will continue to be required over the coming decades.

The Ramsar Convention on Wetlands

The Convention on Wetlands is a treaty between governments that provides an internationally agreed framework for national action and international cooperation for the conservation and wise use of wetlands and their resources. The Convention was agreed in the Iranian city of Ramsar in 1971 and came into full legal effect in 1975. Over the past three decades the Convention has provided a major force for wetland

Ramsar Convention Websites
http://www.ramsar.org

conservation, a major achievement at a time when there is frequent skepticism about the value of international agreements.

Over its short history Ramsar has evolved from its initial emphasis on wetlands as habitats for waterbirds, to focus on the contribution of wetlands to sustainable development. This reflects growing understanding of the multiple value of wetlands and of the need to link conservation to people's needs. As this awareness has grown, so it has driven the conservation investments that can now be seen in many countries.

IUCN-The World Conservation Union

The World Conservation Union plays a unique role in supporting wetland conservation. As an adviser to the Ramsar Convention, and through work with its diverse membership of governments and nongovernmental organizations, the IUCN has helped to identify and pursue some of the most important changes in the international approach to wetland conservation over the course of the past twenty years. These have included early recognition of the importance of linking wetland conservation to the need for sustainable development, helping to establish the World Commission on Dams that has identified more

▶ Fishing on the Logone floodplain, Cameroon. When the river was dammed to provide water for irrigation, the annual flood was greatly reduced and biodiversity and floodplain agriculture all declined. In response to calls from local communities, IUCN brought together a range of partners to develop a system for restoring the flood. This has greatly improved the livelihoods of the community and is one of the best examples of wetland restoration in the tropics.

◀ *Sungai Sedili Besar, Johor, Malaysia. The state of Johor is currently home to 68,630 acres (27,733 hectares) of mangrove forests, out of which about 68 percent are Mangrove Forest Reserves (MFRs). A number of sites in the area have been nominated as Ramsar sites, including Sungai Sedili Besar.*

socially and environmentally sustainable approaches for dam construction, and more recently the launch of a Water and Nature Initiative that highlights the importance of managing water for people and environment.

At field level IUCN works to demonstrate sustainable use of wetlands and water resources, and build capacity to achieve this. Amongst many such achievements, rehabilitation of the Diawling National Park in the delta of the river Senegal in southern Mauritania, and restoration of flooding on the Logone floodplain in northern Cameroon, have provided pioneering approaches to wetland restoration in low income countries.

Worldwide Fund for Nature

Famous for its Panda logo, WWF has also played a major role in wetland conservation from its early days. Amongst its first major conservation achievements was the protection of the Coto Donana in Spain in the early 1960s. Over the past 40 years, both marine and freshwater wetlands have been a high priority for its work worldwide. This has included supporting Ramsar and IUCN, and funding innovative wetland conservation projects in many regions.

WWF Wetlands Website
http://www.panda.org/about_wwf/what_we_do/freshwater

In recent years WWF has expanded its wetland conservation activities through its Living Waters Campaign, and focused increased emphasis on policy engagement. Some major successes have included improving awareness of the economic value of wetlands, continuing to raise awareness of the risks to rivers from dams and other infrastructure, and contributing to the Spanish government's decision to stop the planned diversion of water from the Ebro delta. WWF has also placed major emphasis on establishment of freshwater protecting wetlands, and are working with the Ramsar Convention and other partners to achieve 250 million hectares of protected freshwater wetlands by 2010.

▼ *Sandhill cranes at Lake Andes National Wildlife Refuge, South Dakota, USA. The Lake Andes Wetland Management District, an arm of the US Fish and Wildlife Service was created in the 1960s to protect wetlands in the area from the effects of drainage.*

More information
Information on the work of other organizations working on wetlands conservation can be found at:
Wetlands international www.wetlands.org
Birdlife International www.birdlife.net
The International Crane Foundation www.savingcranes.org
The International Rivers Network www.irn.org

There is a vast amount of information available on wetlands and a quick search on the internet will provide readers with a wealth of sources on different aspects, ranging from recent advances in wetlands science, to current issues in wetlands conservation, and details on many of the organizations and sites described in this guide. To help provide a start some of the main sources of information are given below.

INTERNATIONAL

The most important international perspective on wetlands is provided through the Ramsar Convention at www.ramsar.org (see previous chapter). Information is available on current wetlands conservation issues and access to the Ramsar data base on wetlands of international importance.

Some of the main international organizations working on wetlands and their web sites are:
IUCN – The World Conservation Union
www.iucn.org/themes/wetlands
and *www.waterandnature.org* for information on IUCN's perspective on current wetlands issues and approaches to addressing these. Reports on specific site conservation actions are provided.
WWF
www.panda.org/about_wwf/what_we_do/freshwater for wide range of information on the WWF family's wetlands conservation work.
Wetlands International
www.wetlands.org for information on the organization's work and the Ramsar database on wetlands of international importance.
Birdlife International
www.birdlife.org for information on Birdlife's view of wetlands and endangered wetland birds.
Conservation International
www.conservation.org for information on this NGOs global program which includes work in many large wetland sites such as the Pantanal and the Okavango Delta.
The International Crane Foundation
www.savingcranes.org for the most up to date information on crane conservation, and on many of the wetland sites that they use such as Poyang Lake and the Zambezi delta.

The International Rivers Network
www.irn.org for information on dams and the threats they pose to wetlands and the people that depend upon them.

United Nations Environment Program (UNEP)
www.unep.org/themes/freshwater for links to international monitoring of wetlands and water, including maps of the Aral Sea and Mesopotamian marshes.

United Nations Educational and Scientific Organisation (Unesco)
http://whc.unesco.org for information on the World Heritage Convention and wetlands listed under this Convention such as the Banc d'Arguin in Mauritania and Kakadu in Australia.

The WorldFish Center
www.worldfishcenter.org for information on wetlands and fisheries, and the communities that depend on fishing for their livelihoods.

THE AMERICAS

Ducks Unlimited
www.ducks.org for information on the waterfowl and wetlands conservation activities supported by Ducks Unlimited and its many partners. Links to other wetland conservation actions in the Americas.

US National Parks Service
www.nps.gov for information on the wetlands in the US National Parks system, including at *www.nps.gov/ever* information on the Everglades.

US Fish and Wildlife Service
www.fws.gov for information on wetlands surveys in the United States and on fishing, hunting and other wetland uses. Lots of information on visitor facilities.

Chesapeake Bay
www.chesapeakebay.net for comprehensive information on conservation efforts in the Chesapeake Bay area.

National Wildlife Federation
www.nwf.org for wide range of information on wetlands and their wildlife in the United States. Lots of information on wetlands education.

National Audubon Society
www.audubon.org

The Nature Conservancy
http://nature.org/ for information on the wetlands habitat conservation efforts of this large NGO.

Western Hemisphere Shorebird Reserve Network
www.manomet.org/WHSRN for information on this pan American network of reserves, most of which are wetlands.
Pantanal
www.pantanal.org for dedicated information on the world's largest wetland.

EUROPE
The Royal Society for the Protection of Birds
www.rspb.org.uk for information on the wetlands conservation work of this highly successful NGO, including site conservation, policy, and education.
The Wildfowl and Wetlands Trust
www.wwt.org.uk for information on the work of this NGO whose efforts are focused specifically on wetlands conservation. Good information on visitor sites and waterfowl.
The Wadden Sea Secretariat
www.waddensea-secretariat.org for information on conservation and scientific research in the Wadden Sea.
Lake Hornborga
www.hornborga.com/eng for information on Lake Hornborga and visiting options.
Ligue pour la Protection des Oiseaux (League for the Protection of Birds)
www.lpo.fr for information on wetlands and their birds in France (French language only).
MedWet Initiative
www.medwet.org for information on wetlands conservation work in the Mediterranean.
Coto Donana
www.andalucia.com/environment/protect/donana.htm for information on the Coto Donana including visiting options.
Tour du Valat
www.tourduvalat.org/index_eng.html for information on wetlands conservation research and achievements in France and the wider Mediterranean.

AFRICA
World Heritage Convention
http://whc.unesco.org for information on Convention sites in Africa.
Uganda Wetlands Program
www.ugandawetlands.org for detailed information on

tropical Africa's pioneering wetland conservation program.
South Africa wetlands
www.ngo.grida.no/soesa/nsoer/resource/wetland/

ASIA
Mekong River Commission
www.mrcmekong.org for information on development challenges facing the Mekong and research programs underway, including studies of wetlands and their water requirements.
Wild Bird Society of Japan (WBSJ)
www.birdlife.net/worldwide/national/japan
and *http://www.wing-wbsj.or.jp* for information on birds in Japan, including important wetland sites.

AUSTRALIA AND NEW ZEALAND
Australian wetlands
www.deh.gov.au/water/wetlands for the government's perspective on wetlands conservation in Australia.
Wetland Care Australia
www.wetlandcare.com.au for information on multi-stakeholder wetlands conservation work.
New Zealand Wetlands
www.doc.govt.nz/Conservation/Wetlands/index.asp for information on national wetlands policy.
National Wetland Trust of New Zealand
www.wetlandtrust.org.nz for wider national program of wetland awareness and conservation.

ACKNOWLEDGEMENTS

Many people have contributed information to this guide and to its predecessor "Wetlands in Danger" (Mitchell Beazley 1993). They include for Eastern Canada and Greenland David Boertmann, Steffen Brogger Jensen, and Clayton Rubec; for Western Canada and Alaska Tom Dahl, Clayton Rubec, and Jim Thorsell; for The United States Tom Dahl, Joe Larson, and Dan Scheidt; for Mexico, Central America and the Caribbean Alejandro Yanez Arancibia, Peter Bacon, Monica Herzig, and Enrique Lahmann; for Northern South America and the Amazon Basin Antonio Diegues, Stefan Gorzula, Francisco Rilla, and Glenn Switkes; for Southern South America Argentino Boneto, Charles Duncan, Maria Marconi, Victor Pullido, Francisco Rilla, Hernan Verascheure, and Yerko Vilina; for Northern Europe David Boertmann, Stemar Eldoy, Jens Enemark, Palle Uhd Jepsen, Esko Joutsamo, Torsten Larsson, Karsten Laussen, Hans Meltotie, and Ole Thorup; for West and Central Europe Andrew Craven, Nic Davidson, P. Gatescu, Liz Hopkins, Zbig Karpowicz, Edward Maltby, Francois Sarano, A. Vadineanu, Jurgen Voltz, and Edith Wenger; for The Mediterranean Basin George Catsadorakis, Alain Crivelli, A. Gerakis, Alan Johnson, Thymo Papayannis, Jamie Skinner, and Nergis Yazgan; for The Middle East Andrew Price and Derek Scott; for East Africa Geoff Howard, Paul Matabi, steven Njuguna, and M. A. Zahran; for West and Central Africa Pierre Campredon, Jean-Yves Pirot, and Ibrahima Thiaw; for Southern Africa Geoff Howard, Daniel Jamu, Tabeth Chiuta, and Rob Simmons; for Northern Asia Genady Golubev, Vitaly G. Krivenko, Mike Smart, and Vadim Vinogradov; for Central and South Asia Madan Dey, Zakir Hussain, Peter-John Meynell, Tahir Qureshi, and Sam Samarakoon; for East Asia Derek Scott, Satoshi Kobayashi, and Sara Gavney Moore and colleagues at the International Crane Foundation; for Southeast Asia Zakir Hussain, Faizal Parish, and Marcel Silvius; for Australia Jim Davie, Max Finlayson, and Jim Puckeridge; and for New Zealand and the Pacific, the Department of Conservation (New Zealand), and Derek Scott. For the introductory sections particular thanks are due to Elroy Bos, Edward Maltby, Jean-Yves Pirot, and Jamie Pittock. Any weaknesses in the text however remain the author's responsibility.

PICTURE CREDITS

INDEX

INDEX